Copyright © 2023 by Albert Rutherford. All rights reserved.

No part of this publication may be reproduced, stored in a retrieval system, or transmitted in any form or by any means, electronic, mechanical, photocopying, recording, scanning or otherwise, except as permitted under Section 107 or 108 of the 1976 United States Copyright Act, without the prior written permission of the author.

Limit of Liability/ Disclaimer of Warranty: The author makes no representations or warranties regarding the accuracy or completeness of the contents of this work and specifically disclaims all warranties, including without limitation warranties of fitness for a particular purpose. No warranty may be created or extended by sales or promotional materials. The advice and recipes contained herein may not be suitable for everyone. This work is sold with the understanding that the author is not engaged in rendering medical, legal or other professional advice or services. If professional assistance is required, the services of a competent professional person should be sought. The author shall not be liable for damages arising herefrom. The fact that an individual, organization of website is referred to in this work as a citation and/or potential source of further information does not mean that the author endorses the information the individual, organization to the website may provide or recommendations they/it may make. Further, readers should be aware that Internet websites listed in this work might have changed or disappeared between when this work was written and when it is read.

For general information on the products and services or to obtain technical support, please contact the author.

ISBN: 9798375396880

First Print: 2023 in the United States of America

I have a gift for you...

Thank you for choosing my book, Build a Mathematical Mind! I would like to show my appreciation for the trust you gave me by giving The Art of Asking Powerful Questions – in the World of Systems to you!

In this booklet you will learn:
- what bounded rationality is,
- how to distinguish event- and behavior-level analysis,
- how to find optimal leverage points,
- and how to ask powerful questions using a systems thinking perspective.

Visit www.albertrutherford.com and claim your gift now!

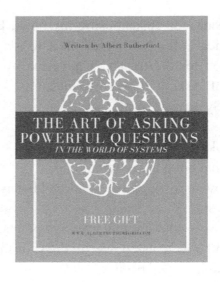

Table of Contents

I Have A Gift For You…   7

Table Of Contents   9

Chapter 1: Mathematical Habits Of Mind   11

Chapter 2: Become A Pattern Detective   25

Chapter 3: Probability And Experimentation   35

Chapter 4: Describing And Speaking In Mathematical Language   45

Chapter 5: Tinkering: Breaking It Down And Putting It Back Together   57

Chapter 6: Inventing: Understanding Algorithms And Using Them   67

Chapter 7: Visualizing: Externalizing The Internal   75

Chapter 8: Guessing: Making Estimations   87

Chapter 9: How Mathematics Changed The World
101

Chapter 10: Final Words 111

Before You Go… 113

About The Author 115

Reference: 117

Endnotes 129

CHAPTER 1: Mathematical Habits of Mind

Ask any adult how they feel about math, and, aside from a few math enthusiasts, you'll get a lot of lukewarm responses. People may say, "Ugh, I hated math," or "I was never good at math! I was much better at reading [or art, or music, or writing, or sports...]," or "It was just so *boring* in high school. My friend and I passed notes the whole time." We have all sorts of reasons for disliking math: maybe we were taught in a "drill and kill" method that bored us to tears; maybe we tried to fit in with a certain crowd in high school by convincing ourselves that we didn't like math (think of Lyndsay Lohan's character in *Mean Girls*). Maybe we even liked math until we got to that infamous train problem in Algebra class. Most of us probably think we aren't very good at math and may have started to believe we weren't "math people" sometime in grade school.

But what is a math person? What if I told you that you could be a math person, too? In fact, anyone can be a math person. This chapter will convince you that you can and should learn to think like a mathematician. The rest of the book will show you how.

Many of us have an idea in our heads of what a "math person" is. Maybe it was the kid in class who raised his or her hand the fastest or the one who always went up to the board to solve proofs in Geometry. Maybe it was the middle-school mathlete or the student who took college-level courses in high school. Sure, one or two of these people may have solved previously unsolved problems, amazing and stunning the world's math community. The rest of them most likely didn't revolutionize the field of mathematics but just enjoyed math during their school years and maybe beyond.

So why did they enjoy math? What habits of mind brought them success in mathematics?

These people knew how to think like a mathematician. Maybe they were born with a predilection towards logical thought, maybe they were trained by talented teachers, or maybe they just enjoyed mathematics enough when they were young they trained their own brains. The point is they learned how to think like a mathematician…and so can you.

Despite what you may have thought in high school, mathematicians have a lot in common with artists, musicians, and other creative thinkers. Mathematics is a creative field that involves visualizing, finding patterns, asking "what if?" and experimenting. What you learned in school – memorizing your times tables or following steps to solve an algebra problem – has little to do with the creative thinking mathematicians do. Many mathematics educators have argued for reforming the way math is taught in school because it has so little to do with what math actually *is*.

In 2009, math teacher Paul Lockhart wrote *A Mathematician's Lament*, a short book that has become a foundational piece for many seeking to reform mathematics education. In his Lament, Lockhart argues mathematics is an art form akin to music or painting, but it hasn't been recognized as such. He faults the educational system, writing, "In fact, if I had to design a mechanism for the express purpose of destroying a child's natural curiosity and love of pattern-making, I couldn't possibly do as good a job as is currently being done— I simply wouldn't have the imagination to come up with the senseless, soul-crushing ideas that constitute contemporary mathematics education."[i]

Lockhart's Lament likens math education to learning to memorize the rules of music in school but never getting to *hear* music until later in life. If we think of the math we learned in grade school as a series of memorizations without getting to experience the art and creativity of doing mathematics, it makes us rethink who a math person is and who isn't. So many of us who were turned off from math at an early age would have loved it had we seen what it truly is. How many people do you know who say "Music? Eh, it's just so boring. I'm not a *music* person."

This is the secret that mathematicians know: math is an art. They know they process mathematics as a musician processes a composition or an artist visualizes a masterpiece. Paul McCartney claims the melody for "Yesterday," one of the Beatles' most beautiful songs, came to him in a dream:

> 'I woke up with a lovely tune in my head,' he told author Barry Miles for the biography *Many Years From Now*, which was published in 1998. 'I thought, 'That's great, I wonder what that is?' There was an upright piano next to me, to the right of the bed by the window. I got out of bed, sat at the piano, found G, found F sharp minor 7th – and that leads you through then to B to E minor, and finally back to E. It all leads forward logically. I liked the melody

a lot but because I'd dreamed it. I couldn't believe I'd written it.'[ii]

Similarly, some mathematicians have claimed that critical mathematics discoveries have come to them while they were sleeping. Srinivasa Ramanujan, an Indian mathematician, believed equations were brought to him in his dreams by a Hindu goddess[iii]; Rene Descartes, the French mathematician after whom the Cartesian coordinate system (our standard way of graphing on two axes) is named, allegedly had his best ideas while lounging in bed in the morning, halfway between sleeping and waking.[iv] Something about the relaxed state of sleeping or being barely awake allowed these people's brains to create, visualize, and dream up ideas related to what their waking minds were focused on.

Recent neurology research has proven math is processed in different parts of the brain than language. A 2016 study by two French neurologists found that people process mathematics in the same parts of the brain where they process problem-solving, which are separate from where language is processed. This can help explain why Einstein allegedly said, "Words and language, whether written or spoken, do not seem to play any part in my thought processes."[v]

Most critically, Amalric and Dehaene, the two French neurologists, found these same parts of the brain are responsible for processing simple mathematics, the kind we learn in grade school.

> The results [of the study] revealed a series of brain areas (from both hemispheres) of the prefrontal cortex, the parietal cortex, and the inferior temporal lobe are activated only when mathematicians are faced with statements or problems of their specialty. And they match the circuits that come into play when anyone handles numbers, does addition and subtraction, or sees a mathematical formula written on paper.[vi]

Mathematicians' brains process advanced math as anyone processes addition or subtraction. This is a revolutionary discovery. It proves we can all be mathematicians or at least think like them. It proves the "innate knowledge that Homo sapiens has of space, time and numbers."[vii]

So, what does this mean for us, the everyday people who are (probably) not about to make a revolutionary math discovery? It doesn't mean we can just go to sleep and expect a discovery to dance into our sleeping minds. You're not going to transition from your everyday life to being a renowned mathematician (or composer) overnight. But you *can* train your brain to think like a mathematician's. That's what this book is about.

Mathematicians' brains are not uniquely formed to solve complex math problems while the rest of us languish amid our basic multiplication tables. When second graders learn math, they are using the same parts of their brains that Ramanujan and Descartes used. Sure, not everyone will grow up to be a Ramanujan or a Descartes, just as not every musician will become the next Paul McCartney, but we all have the necessary parts of our brain to process mathematics.

What is it, then, that sets mathematicians apart? Along with some degree of innate capacity, mathematicians learned to think like mathematicians. They developed the skills they needed to pursue the craft – the art –they loved. Most of them also spent a great deal of time focusing on that art. They spoke with others, they read about new ideas, they tinkered with problems, and they even dreamt about solutions.

Let's delve into that Paul McCartney anecdote a little more. Sir Paul didn't go from learning to pick out notes on the piano to writing the tune for "Yesterday" overnight. He learned the language of music – what the notes on the staff mean, how to read and play chords, what makes a good harmony – and then he thought about it…a lot. His brain was trained to think like a musician's, and we can assume he spent much of his time each day thinking about music.

Researchers and educators have long known training someone to think like a mathematician is possible. How to do that, though, has been a matter of debate. Attempts to reform mathematics education go back as far as your parents or even grandparents can remember (Tom Lehrer wrote the song "New Math" in 1965!). In 1996 (before Lockhart's Lament), a seminal article on mathematical "habits of mind" argued for reforming math education to reflect more accurately what mathematicians do and how they think. The authors open their argument by stating, "Past experience tells us that today's first graders will graduate high school most likely facing problems that do not yet exist." This is even more true now than it was in 1996, before the technological revolution of the 21st century. The authors argue that math education has consisted of memorizing "a bag of facts."[viii]

The authors called for a radical shift in mathematics education, so it focused on the habits of mind mathematicians use rather than the specific facts they have deduced. They proposed teaching students to think rather than teaching them the thoughts that mathematicians have had. They wrote:

> We envision a curriculum that elevates the methods by which mathematics is created and the techniques used by researchers to a status equal to that enjoyed by the results of that research. The goal is not to train large numbers of high school students to be university mathematicians. Rather, it is to help high school students learn and adopt some of the ways that mathematicians think about problems.[ix]

The authors wanted education to focus on creating pattern sniffers, experimenters, describers, tinkerers, inventors, visualizers, conjecturers, and guessers. This was a far cry from the traditional idea of math as basic arithmetic to be memorized.

To clarify what it means to teach mathematical thinking, the Common Core State Standards for Mathematics, first published in 2010, include eight standards for mathematical practice. The SMPs, as math teachers call them, can be taught explicitly alongside math concepts that students learn in their K-12 education. Sometimes, they are written in more kid-friendly language; you may even have seen colorful posters with these practices posted on the walls of your child's classroom. The eight practices are:

1. Make sense of problems and persevere in solving them.
2. Reason abstractly and quantitatively.
3. Construct viable arguments and critique the reasoning of others.
4. Model with mathematics.
5. Use appropriate tools strategically.
6. Attend to precision.
7. Look for and make use of structure.
8. Look for and express regularity in repeated reasoning.[x]

According to these practices, you can think like a mathematician even without formal knowledge. Mathematicians use logic; they look for patterns; they reason abstractly; they don't give up when they encounter hard problems, but rather, they persevere in solving them (or maybe they go to sleep and continue processing them in their dreams). It's easy to see how thinking like a mathematician can benefit us, individually and as a society. These are 21st-century skills.

When we think of math as preparing us for tackling the 21st century's problems, we know we can't take Einstein's words at face value. Despite the fact that language and math are processed in different areas of the brain, we still need language when we're studying math. Maybe you're one of those people turned off by word problems in algebra. I'm not going to tell you that you can tear up the sheets of word problems and throw them out the window. Word problems – specifically, actual real-life problems – are critical to math education.

The problems you saw in math class most likely weren't that relevant to your real life. Maybe you even remember whining to your eighth-grade teacher, "But when am I ever going to need to know this??" Many problems we see in math classes are contrived. You're probably never going to be on a train going a certain direction and need to calculate the exact time it would meet a different train going the opposite direction. But you will encounter all sorts of mathematical problems in your day-to-day life you may not even recognize as mathematics.

Are you a coupon clipper? If you're figuring out your grocery budget and how much you're going to save, you're doing math. Do you look at your cellphone and try to figure out about how much time you have left until you need to plug it in? If so, you're doing math. Have you ever painted the walls of a house or apartment and had to figure out how much paint you needed to buy? That was math.

Sure, these weren't the most exciting math problems– Einstein probably wasn't pondering square footage of walls – but these are real-life problems that involve mathematical thinking. You are probably using math every day in ways you don't even realize. The next step is to identify what it means to think mathematically and to harness those skills so you can tackle more challenging 21st-century problems.

If you ask educators what skills students need for the 21st century, you will get a wide range of answers.[xi] Most people cite technology and the need to sort through what's relevant and irrelevant when presented with a constant stream of information. What is clear is nobody knows what skills we will need in the future, but we're pretty sure they're not the skills we've traditionally learned in school. (So, in many ways, you were right by asking your teacher when you'll ever need to know that!)

In a 2008 speech, the remarkable Sir Ken Robinson likened the education system to factories. We churn out children in "batches" (based on birth year) and expect them all to function identically. This worked well enough when our economy was primarily based on factory work, but our society has changed and continues to change rapidly. "People are trying to work out: how do we educate our children to take their place in the economies of the 21st century?" asked Sir Ken. "How do we do that given that we can't anticipate what the economy will look like at the end of next week?...The problem is that the current system of education was designed and conceived and structured for a different age."[xii]

Sir Ken's answer: schools need to foster divergent thinking or "the ability to see lots of possible answers to a question, lots of possible ways of interpreting a question."[xiii] In other words, we need problem-solving skills. Whose brains are best trained in these skills? That's right: mathematicians'. We know problem-solving skills are needed to address the unknowable challenges coming our way, and we know from neurology research that mathematicians address challenges in the problem-solving regions of their brains.

This means you, the reader, can be uniquely positioned to tackle the challenges of the future if you take a cue from mathematicians. Whatever your background or skillset, you can train your brain by learning how mathematicians think and exercising those skills. In the following chapters, you'll learn to:

1. Develop a mathematical habit
2. Become a better pattern detective

3. Use probability and experimentation
4. Describe and speak in the language of math
5. Tinker
6. Inventing
7. Visualizing
8. Guessing

You may notice these skills are based on the Habits of Mind and the Standards for Mathematical Practice we discussed earlier. That's not a coincidence. Mathematicians and leading mathematics educators know what it takes to think like a mathematician.

In each chapter, you'll learn how mathematicians use each skill or habit of mind, and then you'll find tips and exercises to help you think that way. Don't worry, this is not a math textbook. Rather, it offers a way to train your brain, so you can begin to approach problems with the mindset of a mathematician.

CHAPTER 2: Become a Pattern Detective

Have you noticed, when the weather turns cool, leaves die and fall off the trees? Or cats (and some dogs) like to lie in sunbeams? Of course, you've noticed these things. Humans are born with the propensity to look for patterns. Young children notice, when they cry, a grown-up comes running to comfort them, or if they draw on the walls, the grown-up will get mad. We expect certain behaviors and phenomena because we have observed what usually happens.

So what is a pattern? You might be having flashbacks to math class when your teacher asked you to extend a pattern to find an algebraic rule. (Don't worry, we'll try that later!) But patterns are much more than that; they are everywhere in nature. Look closely at a snail shell, and you will see a pattern. Look at the way water ripples after a stone is thrown, and you will see a pattern. Mathematicians notice and study patterns like these to figure out why they exist and what they can tell us about our world. Sure, they also create patterns for people to examine in algebra class, but the study of patterns exists because people wanted to figure out the world around them. In other words, patterns are a natural phenomenon, and the first step to understanding them is to notice them.

Humans are born with a propensity to look for structure. Neuroscientists have proven through functional MRIs that human brains naturally look for patterns in a sequence of items. In fact, there's a term – apophenia – for "the human tendency to see patterns in meaningless data that may involve visual, auditory, or other senses."[xiv] Our desire to find patterns is so strong we look for them where there aren't any! Extreme pattern-seeking behavior is a hallmark of Obsessive-Compulsive Disorder, Autism Spectrum Disorder, and other conditions. It's much more common to search for patterns obsessively, as people do with these conditions, than it is to not see patterns at all.

"What distinguishes us from most of the animal kingdom is the desire to find structure in the information coming our way," says Robert Barkman, Professor Emeritus of Science and Education at Springfield College in Massachusetts. The neocortex, which accounts for eighty percent of a human brain's weight and is found only in mammals, forms the neural networks responsible for patterns. Humans are so good at pattern recognition that computers have yet to outperform us in that regard.[xv]

The desire and ability to find patterns is an innate skill identified in babies. One of the primary ways babies use their pattern-sniffing skills is through language acquisition. Think about the way you speak. You don't pause between each word so a newcomer to this world (a baby) can recognize where one word ends and the next begins. Studies have shown that babies as young as eight months recognize patterns in speech, such as when certain sounds are grouped together or those moments when we *do* pause between words.[xvi] This pattern recognition helps them make sense of the sounds they hear, leading to the seemingly miraculous moment when they say their first word.

You should be getting the sense by now that pattern-sniffing means more than recognizing the patterns in the tile of a 1950s bathroom. Patterns can be visual, auditory, or "the regular way something happens or is done" (as in how babies learn language).[xvii]

Pattern sniffing, in all ways, is how we learn. Early humans (and animals) learned which plants were safe to eat by watching what happened when their peers ingested different plants. Mothers learn quickly that feeding or changing a crying baby usually stops its crying. Even weather forecasting is based on pattern recognition. While powerful machines now use algorithms to predict the weather several weeks out, early weather forecasting was based on observing what had happened.[xviii]

Mathematicians are adept at finding patterns because they have had years of practice. By slowing down and taking the time to notice, you too can train your brain to notice patterns. Let's try a few exercises to help you become a pattern-sniffer.

Exercise 1:

The first one is simple: The next time you're in a bathroom with some kind of tile or wallpaper pattern, take an extra ten seconds to examine that pattern. How would you describe it? Does it seem to follow a rule? If it's a pattern, it must follow a rule; the key is in describing that rule. Do you see six squares around a hexagon, repeated over and over? Or maybe a print that resets every few inches? Try to imagine this pattern continuing where the tile or wallpaper has ended. Where would each part of the pattern land on the unadorned wall or floor?

Maybe you're more inclined to listen to music than study the wallpaper in a bathroom. The next time you hear a new song, try to pick out the drumbeat or the bassline and see if you can predict when it will change (for example, when the chorus starts). When you're able to make this prediction, you've learned the pattern. If you're someone who listens to or plays a lot of music, you've likely been doing this for most of your life. Your brain is working like a mathematician's without you even realizing it.

Exercise 2:

Let's try another visual pattern. This one is a bit more challenging and gets into some advanced math, though the pattern is quite accessible. Study the sequence below for a few minutes.

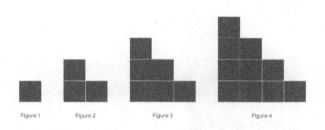

Figure 1 Figure 2 Figure 3 Figure 4

How do you see it growing? Try sketching the next figure in the sequence. Did you follow a rule to make your sketch? What do you think the tenth, fiftieth, or one-hundredth figure in the sequence would look like?

If you were able to describe a rule to extend the pattern above, congratulations! You just dipped your toe in algebra. The pattern above is a classic growing pattern used in algebra courses. Some teachers use it to train their students to notice patterns. Others use it specifically when beginning the study of quadratics. That's right, the pattern above is an example of a quadratic function. The mathematical expression that describes the pattern is: n(n+1)/2. Don't put the book down and run away. That formula simply means the number of tiles (squares) in each figure is the figure number times one more than that number, divided by two. Let's look at Figure 4 for visual proof of that. If you imagine putting two of Figure 4 together to make a rectangle, we can see why the formula works.

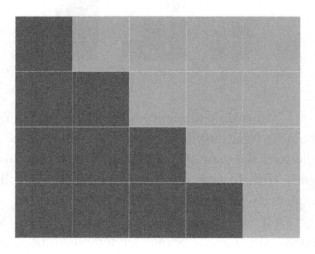

The width of the rectangle is 4 (n, or the figure number); the length is 5 (n+1); multiply 4x5 to make the area of the whole rectangle then divide by 2 to get the staircase figure (made of 10 squares) that makes up half the rectangle.

Not coincidentally, those figures also represent triangular numbers. Let's look at the same pattern written numerically, with each figure represented by a number:

1   3   6   10

How do you notice the numbers growing? What number would come next?

Triangular numbers are numbers that, if represented by dots, would form a triangle. Here's the same sequence shown as triangles:

Triangular numbers aren't just cool to look at. They come up a lot in probability, and they are just one example of figurate numbers, numbers that can be represented by a shape. There are many interesting ways we can examine and use triangular numbers and other figurate numbers, but that's a study for you to pursue on your own time or perhaps in an algebra class. If that exercise didn't pique your interest, don't worry. Just noticing and examining a pattern had you thinking like a mathematician.

Another pattern that abounds in nature can be described by the Fibonacci Sequence. The Fibonacci Sequence is a classic pattern of numbers that begins like this:

0	1	1	2	3	5
8	13	21	34	55	

Can you figure out the relationship between the numbers? Each term in the Fibonacci Sequence is the sum of the previous two terms. After 55 would come 89, since 55+34=89. Fibonacci numbers have been known to Indian mathematicians for over two thousand years. They were first introduced to the western world in 1202 by the Italian mathematician Leonardo of Pisa, who later became known as Fibonacci.[xix]

But why does the Fibonacci Sequence exist? What significance does it have? The numbers in the sequence have a proportional relationship to each other. That means one term divided by the term before gives us a constant (or almost constant) ratio. The longer the sequence continues, the closer the ratio of terms gets to the magical number of phi, 1.618.

The number phi appears so often in nature that it has earned the name the golden ratio. The overlapping spirals in the head of a sunflower almost always contain two consecutive Fibonacci numbers, meaning they exist in the golden ratio. Snail shells contain the golden ratio in the way their size increases as it spirals out.[xx] The way trees branch out follows the golden ratio, with the number of branches growing proportionally as the tree gets taller. The golden ratio is everywhere.

When we look at topics in mathematics that people have studied for thousands of years, it may be difficult to imagine how people got started on that topic. Remember that everything starts with a single person asking a question or noticing a pattern. Someone, at some point, probably hundreds of thousands of years ago, noticed tree branches follow a pattern. Perhaps the same person, or perhaps others, noticed other things around them seemed to follow the same pattern. Eventually, a mathematical theory formed.

To become a better pattern-sniffer, look for and try to make sense of patterns you see around you every day. Once you identify a pattern, visual or auditory or neither, try to describe it. What part of it is repeating? Is there something that changes each time the pattern repeats? What would happen if the pattern continued? You'll soon notice that patterns exist everywhere in our world, inviting you to make sense of them.

CHAPTER 3: Probability and Experimentation

If you've ever played a dice game or watched a movie that involves gambling, you've probably heard of "lucky sevens." You've probably heard the phrase elsewhere too, since it seems to be everywhere. There's a movie called *Lucky 7*; numerous restaurants and bars are called Lucky Sevens; there's a Lucky 7 Casino and even a 2021 film challenge in Las Vegas called the Lucky Sevens Film Challenge ("7 Feature Films. 7 days to shoot. $7,000 budget. The ultimate film challenge!!! Are you in?").[xxi] It seems impossible *not* to have heard of lucky sevens.

Did you ever stop to think about why sevens are supposed to be lucky? It's not random. The "luck" of the number seven can be explained by probability, specifically for games involving dice.

Let's imagine you're playing a game (a gambling game, perhaps) with each player rolling a standard pair of dice on their turn. The dice have six faces, meaning each die has numbers one through six exactly one time. Here's a sample of possible outcomes from a single roll of the dice:

Die 1	Die 2	Total
1	1	2
3	4	7
6	2	8
6	6	12

Here's an exercise for you: find a piece of paper (or open a new document on your computer) and list all the possible outcomes of rolling two dice. Try to work systematically, so you don't miss any. For example, if Die 1 lands on a one, Die 2 could land on a one, two, three, four, five, or six. Record each of those sums separately, since those are all possible rolls.

Once you've worked through the possibilities for Die 1, you don't need to do the same for Die 2, since you've already accounted for each roll. (When you recorded a roll of one of Die 1 and three on Die 2, for example, it's the same as landing three on Die 2 and one on Die 1 – no need to count twice). You should have thirty-six possible outcomes.

Now, look at those outcomes – the sums of the two dice. How many ways are there to roll (as a sum of the two dice) a one? (Zero.) How many ways to roll a two? (One.) A three? (Two.) Keep counting the possible outcomes until you get to seven. How many ways are there to roll a seven? That's right, six. There are more ways to roll a seven than there are to roll any other total.

So are sevens lucky? No! They are just mathematically more probable. Add personal history, culture, and superstition into the mix, and a person might have many reasons for believing sevens are lucky. But in terms of dice, the "luck" of the number seven can be explained by mathematics.

The probability of something happening is always described as a number between zero and one. It's usually described as a fraction or a percent. Zero means absolutely no chance of it happening. What's the probability of rolling a zero when you roll two dice? Zero. You can't roll that.

In the language of probability, one means *certain* to happen. It's rare that something has a probability of one, meaning it's guaranteed to happen. We could say the probability of rolling a sum between 2 and 12 when rolling two dice is 1 (or 100%) because those are the only possible outcomes. But in real life, very few things are 100% guaranteed to happen.

Between those two extremes are fractions that describe how likely a particular outcome is. Probability is calculated by figuring out the number of favorable outcomes – the outcomes we want or are talking about – divided by the number of total possible outcomes. We can write this number as a fraction or a percent.

$$\text{Probability} = \frac{\text{\# of favorable outcomes}}{\text{\# of total possible outcomes}}$$

In the example of rolling two dice, the denominator of the fraction – the total possible outcomes – is 36. The numerator – the number of favorable outcomes or ways that we could roll seven – is six. So the probability of rolling a seven with two standard dice is 6/36, which can be simplified to 1/6.

A little understanding of how probability is measured can help you get through daily life. Computers use advanced algorithms to predict the weather. If the forecast one day calls for a 30 percent chance of rain, it is somewhat likely to rain, and you may want to throw an umbrella in your bag. If there's a 75 percent chance of rain, it is more likely than not to rain. If there's a 95 percent chance of rain, put on your rain boots because you will almost certainly be getting wet that day.

Grain of salt, here: Weather forecasts, as we know from the previous chapter, are derived from patterns of previous weather events. While they have gotten substantially more accurate in the past fifty years, they still aren't correct all the time. If the forecast says 100% chance of rain and it doesn't rain, please don't throw this book at your weather forecaster, but remember weather predictions are just that: predictions.

There's more to probability than understanding how it's measured, though. If you have played a game that involved 36 rolls of the dice, you probably didn't roll *exactly* six sevens, five sixes, and so on. Probability gets a bit messier when we play out scenarios like this in the real world.

In theory, rolling a pair of dice 36 times would give us the outcomes we predicted above, but in practice, our results will most likely be messier. This is the difference between theoretical probability and experimental probability. Theoretical probability is what exists in theory – the pure, raw math. Experimental probability is what happens when we roll those dice.

One fascinating thing about probability is that the experimental probability of an event happening approaches the theoretical probability as the number of trials increases. In layman's terms, this means you're more likely to get those numbers you calculated the more times you roll the dice. If you roll the dice exactly 36 times, you may end up with one seven, five twos, eight sixes, or who knows what else. But if you roll the dice

100 times, 1,000 times, or 1,000,000 times, the fraction of those rolls that are sevens will get closer and closer to 1/6.

This is easier to wrap our heads around when we think about flipping a coin. In theory, the coin should land on heads 50% of the time (or ½ the time) and tails 50% of the time. Most of us have had the experience of flipping a coin several times in a row and having it land the same way. We may have thought we were lucky that day or had a weighted coin. But we probably didn't spend an hour, five hours, or ten hours flipping the coin repeatedly to see what happened. As the hours wore on, the overall outcomes of the coin flips would get closer to 50% each for heads and tails.

What if we flipped two coins, though? Maybe we've placed a bet and we're convinced that flipping two coins will result in one of them landing on heads. If there's a 50% chance of one coin landing heads, wouldn't there be a 100% chance that, with two coins flipped, one of the two will land heads? Probability doesn't work that way, unfortunately.

If you flip two coins, each coin has a 50% chance of landing heads. We can't add those two probabilities because the two coins behave independently. Coin one has a 50% chance of landing heads; so does coin two. We now have four possible outcomes: heads-heads, tails-tails, heads-tails, or tails-heads. The probability of one coin landing heads is still 50% since there are two favorable outcomes out of four possible ones.

Just like noticing patterns, understanding probability will help you understand a lot that goes on in the world. Many aspects of our society are based on probabilities. Think about the health insurance industry in the United States. Before the Affordable Care Act, insurance companies could deny insurance to individuals considered at high risk of costing them a lot of money. These individuals may have had pre-existing conditions or what insurance companies deemed unhealthy lifestyles – a judgment that helped contribute to the discrimination of whole groups of people in the U.S. As any one of these people who were previously denied insurance will tell you, this doesn't mean each of them will necessarily cost insurance companies more. Rather, insurance companies looked at aggregate data to determine how *likely* a person was to incur high healthcare costs.

Data can be examined on an aggregate level, but we approach dangerous territory when we try to draw individual conclusions from it. This is how biases are formed – when people look at aggregate data (whether or not it is correct) and try to apply it to individuals.[xxii]

On a more personal level, understanding probability can help us make better choices in life. Are you someone who buys a lottery ticket every day? Unless you really like to dream (who doesn't?), you may want to stop, as the lottery is a notoriously bad investment. According to Investopedia, the odds of winning the Powerball jackpot are one in 292.2 million. Put another way, it's a .0000000034, or .00000034%, chance you'll win. That's almost 240 times *less* likely than the chance you'll be struck by lightning in a given year. [xxiii] Buying multiple tickets won't help you since the odds of winning are so incredibly slim.

Let's say you're buying raffle tickets, though. If five hundred tickets are sold, and you have one of them, you have a 1/500 chance of winning, or .2%. If you really want that prize, you might want to chance buying a larger number of raffle tickets. If you buy 49 more tickets, so you have 50 and the total sold is 549 (you just added to the total sold, don't forget), your chances of winning are now just over 9%, almost 1 in 10. It's still not a great way to spend your money, since you have a 91% chance of *not* winning, but you do have a better chance of winning this smaller draw than you do of winning the lottery. Plus, the money you've spent on tickets is probably going to a good cause.

The next time you're invited to play a game of chance, or you listen to the weather forecast, or you see a chart of predicted wins in a political race, keep in mind that probability is just that – it can tell us what will *probably* happen or not happen based on patterns and theory, but it can't tell us for sure what will happen.

CHAPTER 4: Describing and Speaking in Mathematical Language

As you've seen, mathematics is all around us. Mathematical phenomena exist in nature. If you take the time to notice and have a little knowledge about what you're looking for, you can use mathematics to understand the world around you.

But there is still something that separates the average person from the mathematician: knowing the language of mathematics. Perhaps the math taught in secondary schools is a language – a second, third, or fourth language for many students. All that time we spend solving equations and deciphering word problems is practice for understanding the mathematical phenomena around us.

Let's look more closely at the language of mathematics. The first thing to understand is that the way we say numbers in English makes less sense than the way they're said in many other languages. If English isn't your first language, you may have struggled to pick up the counting system. Like the customary measurements we use in the U.S. (feet, inches, pounds, etc.), our names for numbers aren't entirely logical.

When we count past ten, we have special numbers for 11 and 12, and then the set of numbers 13-19 tells us how the numbers are composed: thir*teen* means three and ten. Seventeen means seven and ten. Once we get to 20, the order switches, with the number of tens coming first: twenty-one means two tens and one. Eighty-seven means eight tens and seven ones. Romance languages (the most common of which are French, Spanish, Italian, Portuguese, and Romanian) have a similarly irregular set of names for the numbers 11-15 but then switch to the "ten and" structure. Seventeen in Spanish is "diecisiete" – literally, ten and seven. In French, it's "dix-sept" – ten seven.

In Chinese, Japanese, and Korean languages, the structure for counting in teens is much more logical. Eleven is ten-one, twelve is ten-two, all the way through nineteen. Twenty is two-ten; thirty is three-ten; ninety is nine-ten. Ninety-seven is nine-ten-seven.[xxiv] *The way you say it tells you the value of the number.* Think how much easier it would be for young English-speaking children to learn to count if our counting system were this logical!

The English language falls short compared to Japanese again when we look at how we say larger numbers. In English, the number 427, for example, comprises four hundreds, two tens, and seven ones. Students spend a lot of time learning "expanded form" so they can understand the value of each number in our base-ten system. They learn to decompose 427 into 400+20+7. You may remember your second-grade teacher pointing to the 2 and asking what it means. If you said "Two!" like many other children do, your teacher probably shot you an exasperated look, wondering why you didn't know that two actually means twenty.

In Japanese, the number 427 is said as four hundreds; two tens; seven. No decomposing is required because the language conveys the value of each digit.[xxv] Think how much easier it would be to learn to add and subtract if our language conveyed the value of each digit so clearly.

As children get older and move beyond the study of place value, they spend many years learning the language of mathematics, particularly algebra. Were you confused the first time you saw letters in an equation? Until about 6th grade, we were taught that math was about numbers. Suddenly, there are equations with letters in them, and our teachers tell us that x no longer means multiplication but some value we don't know yet. If we've been taught in a procedural way without understanding much of what we do, as many students in the U.S. have been taught math, no wonder we hated it in middle school!

Education reforms have sought to correct many of these misconceptions about math. Place value is emphasized from kindergarten; students now see unknown values in equations from a young age and equations written in different ways (7= ♥ +2). The Common Core State Standards made a serious effort to help children understand the meaning of mathematics. This meant teachers had to shift the way they taught, focusing on meaning rather than procedures.

Algebra has long been known as a "gatekeeper" in mathematics. In New York City, students must pass the Algebra I Regents Exam to graduate from high school. This leads to large percentages of students taking algebra courses once, twice, or even three times. Various studies have shown that algebra pass rates predict high school graduation rates, with some showing those who fail algebra have only a twenty percent chance of graduating high school.[xxvi]

If you're one of those people who didn't pass algebra, don't fear. It's not nearly as scary as you think it is. Algebra is simply the rules of arithmetic you learned previously abstracted. In other words, it takes the rules you already know and generalizes them. For example, look at the simple equation 8-3=5. If we rewrote this using an "unknown" (a variable) for the three, it might look like 8-x=5. Solving for x here just means using what you know about subtraction (or addition, if you wanted to solve it that way). Eight minus some quantity is equal to five. How would you figure out that quantity? Would you count down on your fingers? Add up from five? Any way you solve it is fine because you're demonstrating you understand the structure of subtraction.

Algebra does get more complicated, but even linear equations, such as y=3x+2, are based on arithmetic students learn in elementary school (the four operations of multiplication, division, addition, and subtraction). An equation like this, while it seems abstract, represents a scenario in real life you may deal with all the time. Let's say, for example, you take a cab or a ride-share somewhere. Often, there is a base fare you must pay just for getting in the car – let's say $$2. The driver then charges you $3 per mile you travel. By the end of the ride, you owe the $2 base fare plus $3 for each mile. The x in the equation represents the miles traveled, and the y represents the amount you owe. If you traveled ten miles, you would owe $32 – three times ten plus two.

The reason we use algebra in the above scenario is so we can figure out how much it would cost to ride any number of miles (or our driver knows how much to charge us). The formula is a generalization that allows us to make predictions and see structure. Without it, we might throw a fit when the cab driver announced our fare was $32 because the number would seem arbitrary.

Another way you might use algebra is when figuring out which coupon to use. Bed Bath and Beyond sends out 20% off coupons and $5 off coupons. How do you know which one to use to get the better deal? If you've ever calculated which coupon would save you more money for a purchase, you've thought algebraically.

For those Bed Bath and Beyond coupons, we want to know whether 20% of an item or $5 is the larger amount. The mathematical way to say this is to write two equations: $y=.8x$ and $y=x-5$. It may help your understanding to explain the equations in words, so you can see exactly how they relate to the scenario. The first equation tells us the price you will pay, y, is equal to 80 percent of the original price of the item, x; the second says the price you will pay, y, is equal to the original price of the item, x, minus $5. For the first equation, you may have been thrown off by that .8. We could have written the equation as $y=x-.2x$ (the final cost is the original cost minus twenty percent of the original cost), but we can simplify that to .8x (think about it: if you're not paying for 20% of an item, you are paying for 80% of it, since 100% would be the original amount).

When you solve for a cost of an item – let's say a pillow that costs $18 – you're figuring out which equation will give you a lower value for y:

y=.8(18) (the cost equals eight tenths times 18)
or
y=18-5 (the cost equals 18 minus 5)

Here, the 20% off coupon would give you a final price of $14.40. The $5 off coupon would give you a final price of $13, so it is a better deal for an item that costs $18.

If you wanted to take the Bed Bath and Beyond scenario a step further, you could use algebra to figure out exactly when to use the 20% off coupon and when to use the $5 off coupon. While this seems like complicated math, remember algebra is just the abstraction of rules you already know. It's a language to be learned, but it represents concrete scenarios. We could set those two equations equal to each other to find the price at which they give us an equally good deal. Then we would solve for x, the unknown, the price we're trying to determine.

$.8x = x - 5$

Let's try translating that into regular English first: 80 percent of some amount (x) is equal to that same amount (x) minus 5. Then use the rules of arithmetic you know (or trial and error, picking numbers for x) to solve for x. It turns out $25 is the magic price at which a 20% off coupon and a $5 off coupon yield the same savings. If the price of an item is higher than $25, use the 20% off coupon; if it's lower, use the $5 off coupon. Here's a graph that shows that, where the red line represents the 20% off coupon and the blue line represents the $5 off coupon. The x (horizontal) axis shows the original price of the item, and the y (vertical) axis shows the price after the coupon is applied.

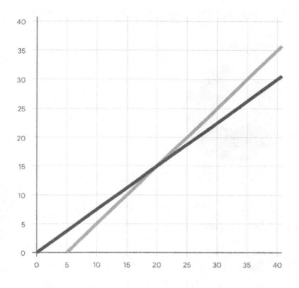

Congratulations! You just studied a system of linear equations! More important, you learned how the language of algebra can help you understand a real-life situation. Just as with studying patterns and probability, understanding the language of math can give you insight into the world around you.

Mathematics is a visual language with symbols and graphs lending information. Think about this when you see infographics in magazines or newspapers. Rather than breezing past those images, ask yourself what information they are trying to convey. Take this one, for example:

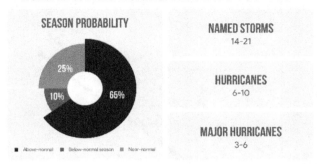

NOAA predicts above-normal 2022 Atlantic Hurricane Season [xxvii]

Study it for a minute to see what it's trying to tell you. The title ("2022 Atlantic Hurricane Season Outlook") and the labels give you necessary information. We know from the label at the bottom that the pie chart is showing us "Season probability." We discussed probabilities, so you've got this! Remember probability is always represented by a number between zero and one, or 0% to 100%. The blue chunk of the pie chart (or circle graph) represents 65% of the chart, and the information at the bottom tells us the chunk shows the probability for an "Above-normal" season. That means it is somewhat likely (65% chance) we will have more hurricanes than usual in the 2022 season. There's a slim chance –

10% -- the season will be "Below-normal" in terms of hurricane frequency.

This infographic conveys important information in the language of mathematics. Understanding it can help you plan for the hurricane season if you live in an area frequently hit by hurricanes. It also lends information to the effort to understand the effects of climate change.

If you found yourself confused by numbers or graphs in school or intimidated by variables and equations, remind yourself that math is a language that represents our world. It is not obscure, created just to perplex generations of students, but a way of communicating information that can help you make more sense of the world around you.

Finally, describing ideas and speaking in mathematical language is important because it enhances our understanding. This seems counterintuitive when you think about a typical classroom. In the traditional classroom, students raise their hands when they think they have an answer; the teacher calls on them and confirms if their answer is correct. Research has shown that this type of instruction, often called I-R-E for Initiate-Respond-Evaluate, doesn't work nearly as well as discussion-based instruction.[xxviii]

In "A Discourse Primer for Science Teachers," published in 2015, the authors lay out five reasons talk is so important in a classroom.

First, and most powerfully, they argue talk is a form of thinking. Linguists and psychologists have proven we don't fully form thoughts before we start thinking; much of our understanding comes from the process of speaking.[xxix] Though we think we know what we want to say, we test, alter, and solidify our thoughts when we speak.

The authors of the *Science Primer* also point out that students' ideas serve as resources for others. Think about a typical discussion. A good discussion isn't each person monologuing their ideas. It consists of different people sharing ideas, building on each other, and coming to a new or shared understanding. In a good math discussion, people may share partially formed ideas, get feedback, and revise their original thinking, eventually landing on a stronger understanding of the topic they're discussing.

Just as mathematicians aren't afraid to experiment, they're not afraid to test ideas before they are formed. They know a discussion will lead to deeper understanding. The next time you have an idea about something, don't be afraid to share it. The process of vocalizing it and getting feedback will only make your idea stronger.

CHAPTER 5: Tinkering: Breaking it Down and Putting it Back Together

When you were a child, did you ever take something apart to see how it works? Even something as simple as a pen or a mechanical pencil can fascinate a curious child or a bored middle-schooler. If you ever did this, you know the value of tinkering with things. Discoveries don't usually come to us as epiphanies but rather as the result of exploration. Mathematicians wouldn't be able to make the discoveries they do without tinkering.

Educators and psychologists have pointed out for nearly a century that people learn by experimenting and trying new things. We construct knowledge rather than having knowledge poured into our brains. Piaget, Dewey, and Montessori contributed to the theory of constructivism, which states children learn by experiencing new things and incorporating them into their existing schema, thereby constructing understanding.[xxx]

Constructivism or experiential education is what mathematicians have long known: phenomena need to be tinkered with, broken down, examined, and then put back together to create knowledge. Tinkering has become more popular in schools, as educators realize it leads to better understanding than memorization does. Many schools now have maker spaces or STEM/STEAM classes, built on the foundation of tinkering and experimentation.

Let's look at a concrete example of how tinkering – breaking things down into smaller pieces and putting them back together – comes up in math class. Elementary school teachers talk about composing and decomposing numbers to help young children gain number sense and an understanding of place value. When adding thirteen and eight, for example, first graders might learn to decompose the thirteen into its pieces -- one ten and three ones. They might tinker with the eight, thinking about ways to decompose it that would make the problem easier for them to solve. Six and two? Five and three? They might realize that seven and one give them friendly numbers to work with: the three ones from the thirteen can be combined with the seven ones to make a new ten. So, thirteen plus eight can be understood as one ten (from the thirteen), another ten (three plus seven), and one one. In other words, twenty-one.

You might be thinking that's a ridiculously complicated way to add two small numbers. Nobody is arguing this is the most *efficient* way to add. Rather, it's a step towards fluency. Educators know memorization without understanding rarely works. Being able to work through this process of decomposition demonstrates and builds fluency with numbers. Students who can break numbers apart like this, looking at their pieces and thinking strategically about how to combine them again, have a much higher chance of succeeding with more sophisticated mathematics.[xxxi] They have a stronger number sense and understanding of place value.

Many concepts in math can be understood by breaking them down. Let's look at a concept that may have scared you in Algebra 1: the distance formula. Many students stumble through Algebra 1, barely understanding what they're doing, memorizing whatever they need to in order to pass tests. The distance formula is one particularly nasty formula that students often balk at (or cry over) when their teachers tell them to memorize it. This formula tells us the distance between two points on the coordinate plane can be found this way:

$$d=\sqrt{(x_1 - x_2)^2 + (y_1 - y_2)^2}$$

You might be asked to use the distance formula to find the distance between points p and q on the graph below, for example:

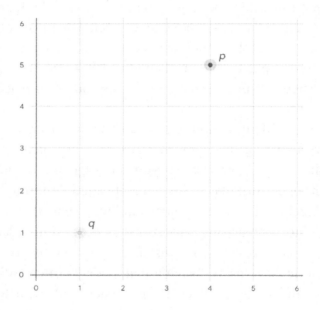

If your eyes just glazed over, don't panic! Mathematicians come up with formulas, so others can follow specific rules and get a reliable outcome (more on that in chapter 6). Teachers teach those formulas in the hopes that they're helping their students get answers, particularly on tests. But the understanding behind formulas like this is what is important, and this formula masks a somewhat simple concept.

To break down this formula, let's recall the Pythagorean Theorem, which you may have learned in eighth grade. The Pythagorean Theorem is a method for finding an unknown side length in a right triangle. (We could spend time building squares and triangles to derive this Theorem, but we'll skip that part, assuming that, at some point you learned and understood it). Let's look at a typical right triangle. We can label the sides a, b, and c. Mathematicians have agreed on conventions in the past; we need to accept that c will always represent the hypotenuse, or the longest side of the triangle, which is always across from the right angle. The other two sides – the legs – are called a and b, and it doesn't matter which one we call *a* and which one we call *b*.

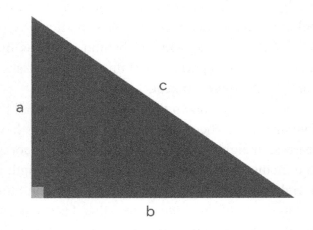

The Pythagorean Theorem tells us the sum of the squares of the two legs is equal to the square of the hypotenuse. The formula looks like this:

$a^2 + b^2 = c^2$

We can use this formula to find the length of the hypotenuse if we have the lengths of the two legs or to find the length of one leg if we have the length of the hypotenuse and the other leg.

Now, let's return to our graph. We can draw a line to connect points p and q, and then make that line the hypotenuse of a triangle (why not?):

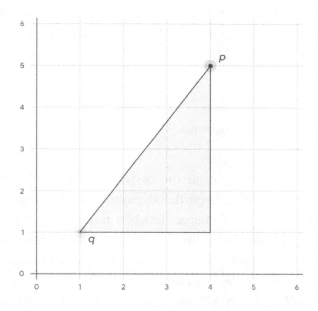

If we know the Pythagorean Theorem, which we do, we can find the distance. We can find the length of the two legs on the graph then use the Pythagorean Theorem to find the length of the hypotenuse. To find the length of each leg, simply count on the graph or subtract the two x-values and the two y-values. In this example, the leg on the bottom of the triangle would be three units (or 4-1), and the vertical leg would be four (5-1). Substitute those into your formula, and you have:

$3^2 + 4^2 = $ hypotenuse^2

To find the hypotenuse (rather than the square of the hypotenuse), we would have to take the square root of both sides of the equation, so we would have:

$$\sqrt{3^2 + 4^2} = \text{hypotenuse}$$

Now, look at the distance formula: that's all it is! We've used the Pythagorean Theorem to figure out the distance between points *p* and *q* is five. If you substituted x and y values into the distance formula for those same points, you would also get five. We can understand what looks like a complicated algebraic formula by drawing a triangle and using what we know about side lengths.

We broke down a formula into something we knew a little better then reconstructed it, creating understanding along the way. That's one way to think about what it means to tinker. Mathematicians tinker by playing around with numbers and formulas, trying things, breaking them apart, and putting them back together, hoping they come to a greater understanding.

Another way to think of tinkering is as a form of experimentation. Tinkerers take ideas, break them into smaller components, and then try things. Will this formula work? What happens if I try this rule? Can I add three to both sides of the equation and get the same thing? These are the questions mathematicians might ask as they tinker and experiment with an idea.

As grown-ups, the rest of us (non-mathematicians) are often hesitant to tinker, particularly with numbers. We've been told that learning, especially in math class, is all about getting the right answer. We are often terrified of doing or saying something wrong in front of others, but we can't expect to learn or make progress if we approach life this way. Every failure teaches us something; we would never learn or achieve anything if we stuck to what we know.

To be a tinkerer, ask yourself, "what if?" Don't be afraid to be extreme! (There's an entire website and multiple books that consist of people asking, "what if?" and coming up with all sorts of wild answers.[xxxii] It's a fun read, particularly if you're in a creative rut!) Anytime you can, try to break ideas (or things!) into their smallest parts, examining and questioning each part. Try different ways of putting ideas back together, changing one small piece to see what effect it has on the whole. Think of the number eight being broken apart in different ways then reconstructed.

Tinkering is especially important if you're trying to accomplish a big task or solve a hefty problem. Try first thinking about the smaller steps you might take to get started. Then experiment. What can you change? What can you try that might help get you to your solution? What new ideas have you found you can incorporate? Don't be afraid to try something and fail. If you've broken down a project into smaller pieces, you've only failed at something small, and that failure has led to learning. Adjust your approach and try again.

CHAPTER 6: Inventing: Understanding Algorithms and Using Them

Do you recall your teachers telling you, "Don't ask why; invert and multiply," or "keep-change-flip"? These are tricks for remembering how to divide fractions. Tricks aren't inherently bad, but they don't tell you how things work. They are memorization tricks that teachers often rely on, believing they're helping their students succeed on future tasks and tests.

Tricks like these often obscure the understanding students need to learn and remember mathematics. There have been several well-known articles and books in recent years that have circulated in the math education community advocating against tricks or artificial "rules" (see *Nix the Tricks*[xxxiii] and "Thirteen Rules that Expire"[xxxiv] for more).

While *memorizing* rules and tricks can obscure learning, *building* rules for yourself can make you more efficient. That's what algorithms in math are all about.

If you have school-age children, you may have found yourself baffled by their math homework. Parents often complain about the way math is taught now, with an emphasis on deconstructing numbers, building understanding, and solving problems in different ways. The Common Core Standards for Mathematics, published in 2012, have been a source of much contention.[xxxv] Parents and some educators have argued we need to return to the traditional way of teaching math, with an emphasis on memorizing and following algorithms. "It worked for us," they often say, "so why change it?"

The problem is it didn't work for us. The American education system has long been known for producing unequal results, with significant racial and economic disparities.[xxxvi] The U.S. has also lagged behind other wealthy nations in international tests of academic progress.[xxxvii] If you're still not convinced, ask the average American on the street to divide two fractions. They will likely try to recite some algorithm they memorized years ago but will misapply it and get the wrong answer. And that's sixth-grade math. Try throwing in a mixed number, and you'll have them stumped!

The Common Core Standards attempt to bring equity to mathematics education and prepare American students for a future yet unknown. They are based on years of research on how people learn. They are based, in part, on the same habits of mind this book strives to teach you – the habits of mind that mathematicians have.

So what is the role of algorithms in the Common Core, or, more broadly, in the life of a mathematician? Let's start by defining algorithm. An algorithm is a procedure for getting a certain outcome. It is a set of steps to be followed that will always work. In math class, "the standard algorithm," refers to how you, your parents, and your grandparents most likely learned to do something. For multi-digit multiplication, for example, the traditional algorithm looks like this[xxxviii]:

```
    4
   36
 × 77
  ───
  252
+2520
 ────
 2772
```

But an algorithm is much more than the traditional way you learned to do something. Remember, an algorithm can be any set of steps that reliably works. If a fourth grader has another method for multiplying multi-digit numbers that reliably works, it's an algorithm. When you balk at your child's math homework, you are likely seeing alternative algorithms you didn't know existed.

The most important aspect of an algorithm is that the person who uses it understands it and can rely on it for solving that type of problem. It is a strategy for making work more efficient. Rather than trying to invent a new way each time, for example, a person can say, "Oh, I know how to do this! I follow these steps."

Algorithms don't appear out of nowhere. The standard algorithm for multiplication (the one illustrated above) exists because, at some point in history, mathematicians decided that was the most efficient way for most people to multiply. Efficiency is based on understanding and being able to follow the steps. Memorization comes easily once someone understands what they're doing and why.

To think like a mathematician, you can't just follow algorithms; you must understand them and invent them. This doesn't mean you need to invent a whole new way to multiply. It means you could spend time thinking about how you do something and why it works then formalizing the steps so you can do that same thing more efficiently next time.

Algorithms apply to much more than the mathematics you learned in school. Algorithms serve as the basis for computer programming and science, for cooking (recipes are algorithms), for advertising, and even for online dating.[xxxix] Online matchups aren't magic (though, if they work, they may seem like it!); computer programs use data to match your specific characteristics with a potential mate's.

You might even use an algorithm when you make a sandwich. You probably have a certain order you follow each time because you know it will be efficient and get you the sandwich you want. Do you smear the peanut butter all over the whole piece of bread, or do you cut it in half first and then smear it on each half? Do you stick the same knife you just used on the peanut butter into the jelly, or do you get a new knife? Whichever method you use, you likely do it the same way each time. You're following your own particular algorithm for making the best peanut butter and jelly sandwich.

So how does one invent algorithms? Inventing algorithms calls for using other habits of mind we've already talked about. First, you need to slow down and observe what is happening. Break the process into smaller, more understandable parts, examining each one. Then put the process back together in an order that makes sense to you and could be repeated another day. When you understand the steps and can follow them again, making your process more efficient, you have an algorithm.

Mathematicians are all about efficiency. There's a saying that mathematicians are lazy.[xl] This means they are always looking for shortcuts. They observe phenomena, look for patterns, and find more efficient ways to do things. They think algorithmically.

One simple way we use algorithms in our everyday life, even when we think we're not using them, is by planning for a change in behavior. Psychologists call this "if-then" planning.[xli] We might think we want to spend less money on eating out, for example. An if-then pattern of thought might be, "If I cook at home five nights a week, then I will spend less money on eating out."

If-then thinking is a powerful psychological tool. It involves thinking about the change you want to make and committing to following the steps to make that change happen. It works precisely because the steps are known and because they are simple or efficient. In this way, it's an algorithm, a formula that can be followed to get a desired result.

Just as a mathematical algorithm makes solving a problem more efficient, if-then thinking takes the stress out of behavioral change. If you convince yourself you are following an algorithm for a desired result, the stress over how to achieve that result dissipates.

Let's look at another example of if-then thinking. Imagine your doctor has said you need to move more every day. Imagine you work on the third floor of an office building. Your plan – the algorithm you're going to follow – could simply be to use the stairs every time you go to your office. *If* you are at work, *then* you will take the stairs, and you'll achieve the desired result: moving more. Whether or not you follow the plan is up to you, but having the plan takes the stress out of trying to figure out how you're going to follow your doctor's orders.[xlii]

Another example of algorithmic thinking that can help you in your daily life is what people call "habit stacking." Habit stacking is when you tack a new desired behavior onto an existing habit to ensure the new behavior will become a habit. As *Esquire* Magazine puts it, it "turns your nagging to-do list into unconscious acts."[xliii]

Let's look at a task that frequently gets ignored and then becomes a source of stress: sorting the mail. Every day, we may get five to twenty pieces of mail, most of which are probably junk. [xliv] It takes most people less than a minute to sort through this mail, yet we procrastinate on doing it until we have a foot-tall stack of mail teetering on the kitchen table and reminding us we need to sort through it.

Instead of ignoring the mail until we can't anymore, we can commit to sorting through it daily alongside something else we do daily, say, taking off our shoes when we walk in the door. A habit-stacking commitment might look like this: *Every day, I will take off my shoes, put them in the closet, grab the mail, and sort through it before I do anything else.* Eventually, sorting the mail will become a daily habit simply because you have "stacked" it with another daily habit you already do.

Habit-stacking is a form of algorithmic thinking. You have a desired outcome – sorting the mail every day – and you come up with a formula, a set of steps that will always work to get you the desired outcome. Following the steps takes the stress out of achieving your goal, since you no longer have to spend mental energy figuring out how and when you're going to do the task. If you follow the steps, you will achieve your goal. That's the whole point of an algorithm – so you can reliably achieve a goal and free up mental energy for other things.

CHAPTER 7: Visualizing: Externalizing the Internal

If you've ever spent time with a gymnast, a concert pianist, or anyone else performing at an elite level, you may have seen them visualizing a performance: imagining every step of it, trying to picture themselves performing it perfectly. You may have wondered why they were doing this rather than spending time practicing leaps or scales. They know something mathematicians also know: visualizing is a powerful tool.

Mathematicians are skilled at visualizing. Einstein attributed his success to the skill: "My particular skill does not lie in calculation," he wrote, "but rather in visualizing effects, possibilities, and consequences."[xlv] It makes sense that visualizing holds such power for humans. Approximately 30% of the brains of primates (which includes humans) is used for visual processing.[xlvi] No wonder we are such visual creatures.

Researchers have identified five aspects of visualizing: internalizing, identifying, comparing, connecting, and sharing.[xlvii] We will examine each of these and discuss how you can hone these skills to incorporate visualizing into your life.

Internalizing involves making sense of something in your head. This is the first step to understanding a problem, particularly a complex one. Let's imagine you're trying to do something that challenges most people: packing a car for a big trip. People who are good at fitting everything into the back of a car aren't magicians; they're just good at internalizing a spatial problem.

When someone gets ready to pack a bunch of suitcases and bags into the trunk of a car, they need to spend time internalizing the problem first. They might ask themselves: How many large suitcases are there? What irregular objects do I need to get in? Are there pockets of space somewhere, maybe under the back seats, where certain items would fit? Where can I put the bag of fragile items, so it's protected and not squished?

The talented packer spends time picturing the answers to these questions and manipulating items in their head before packing the car. If you watch this person in action, you'll see they rarely have to pack and repack the car. They are strategic about what they put where, and they get everything in securely. This is because they spent time internalizing the problem and have a plan for how to solve it.

The identifying stage of visualization involves identifying or creating an image or model that might help you. Young students learn to do this to help them solve math problems. Many teachers use a strategy called, "Read, Draw, Write," which asks students to draw a model or picture to help them solve word problems. They are supposed to read the problem, draw a model, then write the answer in a sentence. The RDW strategy was not created to torture kids or the parents trying to help them with their homework. Rather, it is based on research about how visualization, particularly the act of drawing, creates a stronger understanding and memories of the problem.[xlviii] The process of creating a model leads to better understanding.

Sometimes, we're faced with problems that beg for a drawing to help us solve them. Imagine, for example, you are trying to figure out how to arrange seats at tables for a large party or banquet. You may want to sketch the tables and chairs to help you see the best arrangement. This is a problem that appears in a lot of fourth-grade math curricula when students are learning about division with remainders. These are challenging problems, and teachers often encourage students to draw the scenario or a mathematical model to help them solve it.

Here's another real-life example of how modeling can help you understand a problem. If you've ever shopped at Ikea, you may have used their room planning tool that lets you map out your room to see how the furniture (kitchen cabinets, say) would fit. This software aids your understanding. Without it – without identifying the set-up you need – you would be flying blind. Sure, you can internalize your kitchen cabinet needs, and you can measure, but modeling it can help you tremendously in making a decision. Software like Ikea's exists because engineers know how important modeling or drawing is. Imagine ordering a whole new kitchen without having mapped out if it would actually fit! You'd risk ending up with thousands of dollars of cabinets in your kitchen that needed to be sent back.

Comparing, the next step in visualizing, is a critical piece that often gets omitted when we're short on time. Think back to those kitchen cabinets. Now imagine it's a month before Thanksgiving, and you want a new kitchen in place before your extended family shows up at your house. You might use that kitchen planning tool and then hastily order a bunch of new cabinets without stopping and comparing your model to other possible models. Maybe in three months, you realize your kitchen set-up could have been better had you swapped out one cabinet for a different type or the drop-in sink with an under-mounted sink. In an effort to save time, you skipped the *thinking* part.

Comparing is intricately tied to connecting. There is even an instructional routine called Compare and Connect, where students look at two images or problems and think critically about the differences and similarities between them. The purpose of the routine is "to foster students' meta-awareness as they identify, compare, and contrast different mathematical approaches, representations, concepts, examples, and language."[xlix] A teacher might display two or more models, usually that students have created, and ask the class to compare the models and make connections between them.

A similar math routine is Same But Different, in which students compare and contrast two images. Here's a preschool or kindergarten example[l]:

In this example, young children, who are still learning to count and name groups of objects, might see there are six purple hearts in each image (just being able to count accurately is an accomplishment for many five-year-olds). They might describe the bottom row of hearts as longer. This could lead to a discussion about what it means for a row to be longer. Does that mean the quantity is higher? How can it be that both rows have only six hearts, but one is longer than the other? This is powerful learning for a small child!

Here's an example of Same But Different that one might see in a Geometry class:

This first thing you might notice is that both objects are shapes, but one is yellow and one is purple. Keep looking, and you'll see more. One is a cone, and one is a cylinder. The cone has a circular base; the cylinder has two circular bases (the bottom and top, which can both be called bases). You might think about how the volume or surface area of the two shapes compares. You might wonder if the cylinder could hold more liquid than the cone and how much more. In just a minute or two, by comparing and contrasting these two images, you thought about deep mathematical concepts involving spatial reasoning. You also used specialized language to discuss your ideas.

That brings us to the final aspect of visualizing: sharing. Using language to describe what is in your head solidifies it. Think back to the chapter on using mathematical language and remember that talking is a form of thinking. It also makes your thinking available to others, so they can learn from it.

Let's return to our kitchen remodeling scenario. If you took the time to compare and make connections between the different kitchen set-ups you might go with, you would have your ideal kitchen. Now imagine describing or showing that kitchen layout to a family member. While describing where the kitchen island is, you may realize you made a mistake on a measurement and want to revisit it. Your relative may point out that you forgot a spot for storing pot lids, which you wouldn't have realized if you hadn't shared your plan. Not only will you be checking your work, but you'll also be helping your family member understand the process of planning a new kitchen, should they need to do it someday.

We looked at a couple of examples of how visualizing can help you in everyday life. These examples included spatial problems – how to fit luggage in a car and how to lay out a new kitchen. But visualizing can help you with much more than spatial problems. It can help you plan for the future and be prepared to tackle the problems you might encounter.

Just as if-then planning and habit stacking can be powerful tools, psychologists and neuroscientists have recognized the power of visualizing. Muhammad Ali famously said, "If my mind can conceive it and my heart can believe it – then I can achieve it."[li] Athletes have used visualization to decrease performance anxiety and to increase performance results. Musicians sometimes review how to play a piece in their head, imagining flawless finger positions as they go along. Studies have shown that visualizing what you want to achieve – say, a perfect tennis stroke or a perfectly played concerto – leads to better performance. If the brain sees it, the brain is more likely to do it.

This seems like a trick; how can visualizing something make you better at doing it? It has been proven time and again; it actually works! A recent study looked at a small-scale version of this: could a group of study participants increase the strength in their pinkies just by visualizing pinkie exercises? The study proved those who performed "mental contractions" of their little fingers did gain muscle strength! They didn't gain as much as the participants who performed actual pinkie exercises, but they gained noticeably more than the control group, which performed neither mental nor physical exercises. The study concluded "that the mental training employed by this study enhances the cortical output signal, which drives the muscles to a higher activation level and increases strength."[lii]

This tells us that visualizing can and should be regularly practiced. Anyone can benefit from visualizing, not just elite athletes or musicians. Do you have an upcoming dentist's appointment you're worried about? Picture yourself sitting calmly in the chair, unperturbed by the drilling in your mouth. Planning for a difficult conversation with your boss? Imagine yourself calmly walking into his or her office, clothes unruffled, tackling the conversation with confidence. These visualization practices will not only help you feel less nervous, but the practice of visualizing is more likely to make the event happen the way you want it to.

The five steps of visualization we discussed earlier are useful for tackling complex problems. Maybe you don't know what you want to say to your boss, but you know you need to have a conversation. First, try to think about what you want the conversation to be. What is the main issue you want to discuss? How do you want to come across? What points would help make your position stronger? Then "model" the conversation. Maybe this means having a practice conversation with a friend or writing down what you want to say. Compare it to other tactics or arguments you could make and connect it to past conversations you've had with your boss. This can help you see any points you may have omitted or plan a rebuttal for the argument your boss may pose. Finally, share your plan with someone else. The act of sharing will solidify the plan in your head and give you a chance to hear feedback. Repeat this process as often as you need to once you've gotten feedback. Now you should be prepared to talk to your boss.

Another process that can help you achieve things you want in your life is making a vision board. A vision board, often used to map out what you want your life to look like in a few years, can encompass each of the five steps of visualization.

First, internalize what you want your life to look like. Then identify what those aspects of your life might look like. Find images to represent what is in your head. Then compare and connect these images. Are there others that might work as well? What do these images have in common? What do they reveal about your goals? Finally, share the vision board, either with yourself (display it somewhere you will be reminded of it) or with a friend. This helps cement it in your mind.

Now that you've learned the benefits of visualizing, you are truly on your way to thinking like a mathematician. There is one final skill to discuss: guessing and making estimations.

CHAPTER 8: Guessing: Making Estimations

The final habit of mind is one you use every day, whether you realize it or not: making estimations. Every time you go to the grocery store (unless you bring a calculator with you), you probably estimate your spending. When you need a tank of gas, you may estimate how much it's going to cost. Before leaving the house, you may grab a twenty-dollar bill, estimating that anything you need money for will cost less than that. Before leaving for work in the morning, you estimate how much time you'll need to shower and get ready.

Estimating, or making educated guesses about numbers, comes up all the time in everyday life. You may find some situations easier to estimate than others, and you may have friends that seem better than you at estimating. Estimating involves number sense, which is a person's ability to understand and manipulate quantities. Some of us have stronger number sense than others.

Educators have realized how important number sense is, as it forms the foundation not only for estimating, but for grasping numerical and spatial concepts easily. Much of math class now focuses on building students' number sense, particularly in the youngest grades. If you have heard of students doing "number talks" in class, know these students are building number sense. The stronger their number sense, the better their ability to estimate will be.

Let's look at how you use estimation in your everyday life. You use it not just for costs, as the previous examples illustrated, but also for all kinds of quantitative scenarios. Imagine you're walking into a museum or large office building, and you see a big set of stairs ahead of you. Without realizing it, you mentally approximate how many steps there are and therefore how much effort you'll need to exert before deciding whether to look for the elevator. If there are only a few stairs in front of you, you'll probably take them. If it's a hefty flight, you may decide you'd be better off with the elevator.

Now imagine you're about to go for a long drive, maybe a once-yearly trip back home to see relatives. The map on your phone can tell you approximately how long it should take – let's say five hours. Because you have driven this route before, you know when and where you might encounter traffic. You might estimate an additional two hours if you're driving during rush hour. Maybe you will have a toddler in your car, and you'll tack on another hour for planned bathroom breaks. When you talk to your family on the phone the night before, you may tell them you plan to leave at 11 and get there at 7 because you've estimated those additional three hours.

This scenario demonstrates that personal experience also plays a role in estimating well.[liii] An experienced salesperson might estimate their profit on a new product before it's available to the public. A kindergarten teacher can predict how long it'll take twenty-five five-year-olds to walk from the playground back to their classroom. A good knitter can estimate how long much longer it'll take to knit a patterned sweater than a simple hat. An experienced chef knows what a pinch of salt or approximately a teaspoon of something looks like. These estimates are based not just on knowledge of quantity but on personal experience.

While you can't feign experience you haven't had, you can improve your number sense. One of the first things young students learn is what many math programs call "friendly numbers" or benchmark numbers. In our base ten system, friendly numbers are usually multiples of ten or one hundred. Friendly numbers are much easier to work with than other – let's say non-friendly – numbers.

Young children learn about friendly numbers when they are first learning to add and subtract. Let's look at the problem 18+7 as an example. Adding eighteen and seven is typically difficult for a first grader. Instead of wracking their brains tackling 18+7, students might be instructed to find the nearest friendly number to the first addend – in this case, twenty. To make twenty, we would need to add two to eighteen. We can then compensate by taking two away from seven. Then our math problem is 20+5, which is much easier to solve in our heads than 18+7.

Now, let's look at an example an adult might encounter. Have you gone out to a meal with someone who always leaves a whole-dollar amount for the server? Some people use a tip calculator or their phone to calculate the tip, but others use their number sense and friendly numbers. Imagine your check is $38.46, and you want to leave a tip somewhere around 15%, maybe a little more. Ten percent of $38.46 would be $3.84; twenty percent would be double that, or $7.68. Fifteen percent would be halfway between those two numbers -- $5.77, to be exact. Without a calculator, though, many people wouldn't figure out that exact percentage. They'd be much more likely to think the tip should be somewhere between $4 and $8 and use a friendly number to decide.

With the $38.46 check, your friend might leave $45 and calculate the exact tip after they have made that decision. For most people, it's easier to figure out the difference between $45 and $38.46 than it is to calculate fifteen percent, then add that fifteen percent ($5.77) to $38.46 in their head. This is because $45 is a friendly number, since it has zero cents attached to it. Starting from $38.46, we can count up 54 cents to the next whole dollar, essentially counting up to the next 100 (since 100 cents makes a dollar). This gives us $39, and then we can add six more dollars to get to $45. We've estimated the tip – somewhere between ten percent and twenty percent – and used a friendly number to figure it out exactly.

Using friendly numbers will get you far in life. It's an essential skill for doing mental math. It is closely related to, but not quite the same as, the ability to round numbers. When we use friendly numbers, we keep track of where we need to compensate, and we find an exact answer. When we round, we're trying to get a general idea of a quantity.

Children learn how to round in elementary school, usually focusing on it in third or fourth grade. It is something they will use almost every day of their adult life. When they bug their teachers with, "But will we ever need to know this?" their teachers can resoundingly answer, "Yes!" Rounding means turning the number into the closest group of whatever you're asked to round to. Fifty-four rounded to the nearest ten is fifty since fifty-four is closer to fifty than it is to sixty. Rounded to the nearest hundred, it's one hundred since it's closer to one hundred than it is to zero.

You probably use rounding most often in dealing with money, and this is where it can be most useful. Retailers like to make prices seem lower by taking one or two cents off. You're more likely to see meat advertised at $2.99 a pound than at $3 a pound. Market researchers have found that consumers aren't that savvy.[liv] Because we read from left to right, we're likely to focus on the two in $2.99 and think "two dollars a pound! That's a good price!" But, if we use our number sense and round to the nearest whole dollar, we know $2.99 is basically $3.00. Next time you go shopping, remember not to be fooled by marketing strategies but to use your number sense!

Rounding helps us in all sorts of everyday scenarios. While you're shopping – hopefully after you've realized the meat is about $3 a pound, not $2 a pound – you may round to keep a running tally of how much you are spending. About $5 for that box of granola bars, $3 for that bunch of vegetables, $4 for the carton of milk... If you're budgeting or watching what you're spending, this is probably a familiar scenario.

Teenagers do this all the time since their parents are usually the ones managing their money. A teen might get $5 to spend at the corner store on their way to or from school. They may want chips and a drink. Before they get to the cashier, they're probably rounding and adding approximate amounts in their heads. They don't want to be caught short, and most teens aren't going to expend the time or effort to add, say, $2.37 and $1.98 in their heads. Adding $2.50 and $2.00 is much easier, and they'll know they still have some wiggle room, maybe for a piece of candy by the register, since they rounded up both numbers.

Builders and contractors also round and estimate. Underestimating a cost will lead to slow-downs and upset the client, so they will round up whenever they can. If they calculate that they need just over seventeen boxes of tiles to complete a floor, they'll round up and order at least eighteen. Some contractors default to ordering ten percent more of whatever product they need, so they don't end up short. They give the customer an estimated price and amount of time to complete the work, basing their estimate on their experience doing similar projects. After their initial assessment and set of measurements, their ability to get work depends on estimates.

We also round when we talk about time. If your partner or roommate asks how long until you're ready to go, you probably won't respond "twenty-seven minutes and forty-nine seconds." We round amounts of time because a) it's easier for us to say, and b) we know the person we're talking to will understand the rounded amounts. Their lived experience and sense of time give them an idea of what approximately half an hour feels like. It also gives us some leeway. If we're ready in 25 minutes, great. If it's closer to 35 or even forty minutes, our roommate probably won't care.

So how can you improve your ability to estimate? The first thing you can do is try solving problems in your head whenever you can – the mental equivalent of taking the stairs instead of the elevator. If you're given a check at a restaurant, spend a few minutes thinking about what ten percent of the amount might be then twenty percent. Use these numbers to figure out the tip, or use the friendly number strategy to determine the tip after you've decided what the total should be. At the grocery store, keep a running tally of your items, and make sure you round accurately, figuring out what whole-dollar amount the item's real price is closest to. When it matters, though, don't be afraid to use a calculator – knowing how and when to use appropriate tools is also an important skill and one of the practice standards in the Common Core State Standards for Mathematics.

We'll end this book with an exercise that can strengthen your number sense and is one of the most powerful tools a teacher can use in the classroom: a number talk. This won't really be a number *talk*, since you're reading, but it'll be a chance to think about a numerical expression in different ways. This particular one is from a blog on teaching mathematics.[iv] You'll see an image, and your job is to figure out how many dots you see:

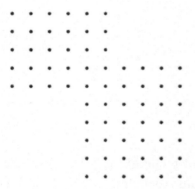

Think of as many ways as you can to figure out how many dots there are in the image without counting them one by one. This is where the *talk* would come in. A teacher could solicit several explanations and have students compare them. So, since this is a book, we'll list and describe them, and you can try to draw the connections.

Let's get the answer out of the way first. There are 68 dots. If you got it wrong, don't worry. Part of the goal of a number talk is to see different strategies for solving (or counting) and where you went wrong. Here's one strategy:

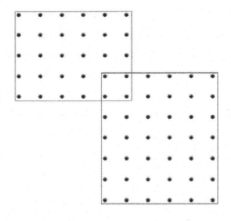

The red rectangle is 5 by 6, so it contains 30 dots. The blue rectangle is 6 by 7, so it contains 42 dots. But the two rectangles have four dots that overlap in the middle, and we don't want to count those dots twice, so we have to subtract four. Our count could be represented by this equation:

$(5 x 6) + (6 x 7) - 4 = 68$

Or perhaps you broke the image up into three rectangles. There are a number of ways to do this; here is one:

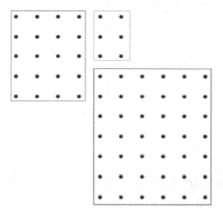

In this image, we have a rectangle made of 20 dots (red), one made of 6 dots (green), and one made of 42 dots (blue). Our equation could be:

$(4x5) + (2x3) + (6x7) = 68$

Another way to break the image into three rectangles is like this:

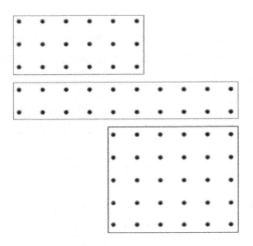

An equation to represent this breakdown would be:

(3x6) + (2x10) + (5x6) = 68

Or maybe you thought outside the box (no pun intended) and pictured the shape that would go around this dot image then subtracted the number of dots you imagined would be in the negative space:

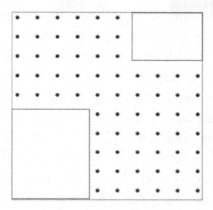

This would give you the equation:

$(10 x 10) - (3 x 4) - (4 x 5) = 68$

The next question in a number talk, and where the real mathematical thinking lies, is why do all these equations work? They look different; how can they all be equal? What mathematical rules underlie our equations that allow us to write them in such different ways? What similarities and differences do you see among the equations?

In that one number talk, we demonstrated most of the mathematical habits of mind. We looked for a pattern, noticing the arrays the dots form; we experimented with different ways of counting, possibly making mistakes along the way; we described what we saw in mathematical language; we tinkered and broke the image down into different components; we came up with strategies or algorithms to help us solve; we visualized the different ideas that were in our heads; we compared and connected the different strategies (we would have done this more had we been talking, not reading!). We didn't need to estimate because we could find an exact number, although we could have estimated at the beginning, before proceeding with equations. A good number talk incorporates all these habits and so much more.

A number talk can be a discussion about numbers or an image that can be broken down in different ways, reinforcing mathematical rules. Challenge yourself to see numbers and patterns in different ways, looking for new strategies and connections. Thinking through tasks like this will increase your number sense and open your eyes to the beauty of mathematics, which can be represented in many different ways.

CHAPTER 9: How Mathematics Changed the World

If you're still feeling skeptical about the value that mathematics can bring to your life, perhaps a short lesson in history will help convince you how powerful it can be. History has a funny way of being rewritten and manipulated, depending on who's telling the story and what purpose they want the historical information to serve. We rarely hear about how mathematicians changed the world, but it's true. There are numerous examples throughout history of the incredible contributions mathematicians made to civilization.

Let's start with the ancient Sumerians, one of the earliest known civilizations in the world, often credited with being the "cradle of civilization." The Sumerians lived in Mesopotamia in the area that is modern-day Iraq and flourished from approximately 5000 BCE to 2000 BCE. Sumerian civilization bloomed, in part, because humans learned to cultivate farmland, which lead to increased food supply, enabling population growth and the establishment of large population centers (city-states).

What does a growing population center need to amass wealth? Math. Specifically, the Sumerians needed a numbering system and basic calculations to help them keep track of land and taxes. Ancient clay tablets from the Sumerian city of Ur give us evidence of how King Shulgi, who ruled over Ur from approximately 2094-2046 BCE, created the first "mathematical state."[lvi] Shulgi had hymns written about his prowess in nearly everything (and even declared himself a god during his reign), so we can't be sure how much of a mathematical genius he was. He did, however, standardize weights and measures, a crucial step in keeping track of state finances.[lvii] (Can you imagine trying to rule an empire *without* standardized weights and measures?)

The Sumerians also created one of the first numbering systems (or one of the first we have evidence of, thanks to those clay tablets). Look at the numbers one through fifty-nine written in cuneiform below[lviii]:

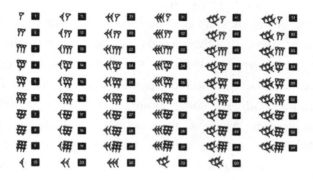

Practice your ability to notice patterns and examine the table above for a minute. What do you see? Do you notice how the number eleven is the symbol for ten next to the symbol for one? And the pattern continues all the way through fifty-nine? It's similar to the way we write numbers, using place value to help us write numbers greater than nine. One crucial difference between their system and ours is that, once they got to sixty, they restarted with the same symbol used for one and then would use place value to create higher numbers from there.[lix] Remnants of their base-60 system still exist today, for example, in how we tell time.

There was a small problem in the Sumerians' system, though. The numbers one and sixty were represented with the same symbol;

there was no symbol for zero to indicate sixty was one group of sixty and nothing in the place to the right of it. In other words, without a zero, there was no way to indicate the difference between numbers that relied on position, or place, to be understood. At some point, perhaps in ancient India[lx] or perhaps in multiple civilizations at different times, a symbol was used to represent nothing (zero) in a position.[lxi] It may seem trivial, but without the symbol for *nothing*, people's ability to indicate large quantities was limited. Once a symbol for zero came into existence, our capacity to represent numbers using place value became, well, infinite.

Ancient civilizations, including Indian, Mayan, and Egyptian civilizations, came up with all sorts of fascinating mathematics that influenced later cultures. It might be more appropriate to say they didn't come up with the mathematics, but they noticed and harnessed the power of mathematics, figuring out ways to notate it and use it to their advantage. After all, mathematics is the study of phenomena that exist around us; it's our job to notice and interpret those phenomena. If you look at cultures all over the world throughout history, you'll find countless instances of mathematical "discoveries" that came about because someone needed mathematics to solve a problem.

Pythagoras is one of the most well-known ancient mathematicians for good reason. The Pythagorean Theorem (which you learned about in Chapter 5) helps us understand basic geometry and leads us to more advanced mathematics. It is controversial, though, whether Pythagoras discovered the Theorem. Evidence suggests that knowledge of the Theorem –the sum of the squares of the two sides of a right triangle equals the square of the hypotenuse – existed in ancient India and perhaps elsewhere, and Pythagoras may have learned of it during his travels.[lxii] In fact, Pythagoras founded a school of philosophy (known as the Pythagoreans) and was credited with many things that may have been proven or discovered by others. Whatever the source, though, the Theorem was a critical discovery that allowed for more discoveries in mathematics, physics, art, architecture, topography, and other fields. It is one of the crucial building blocks of civilization.

The history that many residents of the Western hemisphere learn traces a line from ancient Babylonia to ancient Greece to Renaissance Europe, focusing entirely, or almost entirely, on Western mathematicians. Many non-Western cultures contributed to mathematics before or contemporaneously to the Europeans, but not as much has been written about them.

Some of these mathematical traditions exist orally or in other forms (such as music and art), making them harder for outside cultures to learn about. More sources are becoming available every year, though; a decade from now, we may have a new book to write about how non-Western mathematical discoveries changed the world.

One massive contribution between the Greeks and Renaissance Europe that Western students do learn about is the field of algebra. While aspects of algebra had been used for centuries in some cultures, Muhammad ibn Musa al-Khwarizmi, a 9th century Muslim mathematician and astronomer, defined it. Khwarizmi's comprehensive book, titled *The Compendious Book on Calculation by Completion and Balancing*, formalized the processes for solving linear and quadratic equations. He also came up with the idea for algorithms, paving the way for other mathematicians to formalize their processes.[lxiii]

It's not a stretch to say algebra exists as its own topic we study in school because of Khwarizmi's work. If you hated algebra, though, you shouldn't resent Khwarizmi. Algebra is simply the formalization and abstraction of operations and rules you learned previously. Khwarizmi recognized this and gave it language, so others could use it when they needed to.

Now, let's jump ahead to 17th-century Europe, where German astronomer Johannes Kepler paved the way for an entirely new field of mathematics because he wanted to save money on wine. The story goes that Kepler realized the current method for measuring how much wine a merchant sold was faulty, leaving customers at the mercy of the wine merchant, who might be overcharging them. Kepler wanted to figure out an exact way of measuring how much wine was in a barrel, so he created a method for figuring out the volume inside a curved shape (a wine barrel). He then published a book about his findings, *The New Solid Geometry of Wine Barrels*. Mathematicians see this book as a foundational text in integral calculus, with later mathematicians drawing on Kepler's work.[lxiv]

While most people don't use calculus in their everyday lives, calculus is a critical component of many fields, including engineering and medicine. Calculus is used to figure out minimums and maximums in many fields. It is even used to calculate minimum credit card payments.[lxv] The next time you make a credit card payment or buy a bottle of wine for a reasonable price, thank mathematics!

There's an even more ubiquitous time you could be thanking mathematics: anytime you turn on a light, watch tv, listen to music in your home, use a computer, or do anything else that uses electricity. You probably know of Thomas Edison as the person to invent the lightbulb, but you likely haven't heard of the man responsible for bringing electricity to households around the country: Charles Proteus Steinmetz. Steinmetz, born Karl August Rudolf Steinmetz in what is now Poland, was a mathematician whose mathematical discovery helped create electrical circuits, the critical component bringing electricity wherever we need it.[lxvi]

Steinmetz's discovery involved a topic that may bring back memories (or nightmares) from your high school years: imaginary numbers. Mathematicians had, for centuries, been stumped by what seemed like a mathematically impossible situation: no number multiplied by itself will ever give you a negative number; thus the square root of a negative number seems impossible. But there had to be some way to define the square root of a negative number (or, more specifically, negative one) since some mathematical operations lead to a solution that includes this.

Mathematicians called these square roots "imaginary numbers" but didn't really have a use for them. That was until Steinmetz came along. Steinmetz figured out how to simplify complex formulas for electrical circuits using imaginary numbers, thus enabling electrical circuits to be more easily and widely produced.[lxvii] Without Steinmetz's mathematical discoveries, we might be – quite literally – in the dark.

Mathematical formulas are also at the heart of many scientific discoveries that changed the course of history. Isaac Newton's Law of Universal Gravitation, Einstein's Theory of Relativity, the Second Law of Thermodynamics, and Chaos Theory all involve complex mathematics.[lxviii]

Mathematics is at the heart of civilization in every culture around the globe. Ancient civilizations needed mathematics as their societies developed, initial mathematical discoveries led to more complex ones, and now almost every aspect of modern life can be traced back to a mathematical discovery or understanding. So the next time you hear a teenager whining about when they'll ever need to know the math they're learning, you can tell them what you know: math is the foundation of everything. Our civilization wouldn't be where it is now if it hadn't been for talented thinkers who noticed mathematical phenomena and then harnessed its power for human advancement. Math forms the basis of so many professions and so much of our daily lives.

CHAPTER 10: Final Words

You have learned the habits of mind that mathematicians rely on and that underlie modern mathematics instruction. You've learned to sniff out patterns, understand and use probability, speak in the language of mathematics, tinker, invent, visualize, and make educated guesses. You've also learned, if you didn't know it before, that mathematics shaped our history and plays a role in nearly every aspect of life.

One of the key differences between mathematicians and everybody else is that mathematicians don't give up when they face a mathematical challenge. In fact, they enjoy these challenges. They know learning comes from perseverance, mistakes lead to greater knowledge, and even seemingly insurmountable problems may have solutions. In short, they believe in themselves and their problem-solving abilities.

The other thing that mathematicians understand is that math makes sense. It's not an obscure topic designed to torture students and stump grown-ups. Math is logical, and it can be understood by anyone who takes the time to try to understand it. Once you begin to unlock puzzles that seemed impossible to you, you will see that you *can* do it. You too can make sense of complex mathematics and begin to harness the power of mathematics in your life.

Whether you're an artist, an engineer, a mechanic, a bartender, a professor, a teacher, or anything else, the mathematical habits of mind you learned in this book can help you as they helped centuries of thinkers before you. Try to keep them in the forefront of your mind and draw on the habits you need at different times. This is how to think like a mathematician – be able to pick the right tool and discern the most efficient way to solve a problem.

Respectfully,

A.R.

Before You Go...

I would be so very grateful if you would take a few seconds and rate or review this book on Amazon! Reviews – testimonials of your experience - are critical to an author's livelihood. While reviews are surprisingly hard to come by, they provide the life blood for me being able to stay in business and dedicate myself to the thing I love the most, writing.

If this book helped, touched, or spoke to you in any way, please leave me a review and give me your honest feedback.

Visit Amazon.com to leave a review.

Thank you so much for reading this book!

About the Author

Albert Rutherford

We often have blind spots for the reasons that cause problems in our lives. We try to fix our issues based on assumptions, false analysis, and mistaken deductions. These create misunderstanding, anxiety, and frustration in our personal and work relationships.

Resist jumping to conclusions prematurely. Evaluate information correctly and consistently to make better decisions. Systems and critical thinking skills help you become proficient in collecting and assessing data, as well as creating impactful solutions in any context.

Albert Rutherford dedicated his entire life to find the best, evidence-based practices for optimal decision-making. His personal mantra is, "ask better questions to find more accurate answers and draw more profound insights."

In his free time, Rutherford likes to keep himself busy with one of his long-cherished dreams - becoming an author. In his free time, he loves spending time with his family, reading the newest science reports, fishing, and pretending he knows a thing or two about wine. He firmly believes in Benjamin Franklin's words, "An investment in knowledge always pays the best interest."

Read more books from Albert Rutherford:
[Advanced Thinking Skills](#)
[The Systems Thinker Series](#)
[Game Theory Series](#)
[Critical Thinking Skills](#)

Reference:

A Discourse Primer for Science Teachers. (2015). In *Ambitious Science Teaching.*

Achievethecore.org : (n.d.). https://achievethecore.org/content/upload/Mathematical+Routines.pdf

Al Jazeera. (2015, October 20). *Al-Khwarizmi: The Father of Algebra.* Science and Technology | Al Jazeera. https://www.aljazeera.com/program/science-in-a-golden-age/2015/10/20/al-khwarizmi-the-father-of-algebra

Algorithms in Mathematics and Beyond. (2018, December 21). ThoughtCo. https://www.thoughtco.com/definition-of-algorithm-2312354
Authors.

Barkman, PhD, R. C. (2021, May 19). Why the Human Brain is So Good at Detecting Patterns. *Psychology Today: A Singular Perspective.* https://www.psychologytoday.com/us/blog/singular-perspective/202105/why-the-human-brain-is-so-good-detecting-patterns

Buell, B. (2017, June 18). 2 wizards kept GE atop technology field. *The Daily Gazette*. https://dailygazette.com/2017/06/17/edison-steinmetz-forged-long-friendship/

Cardone, T. (2015). *Nix the Tricks: A Guide to Avoiding Shortcuts That Cut Out Math Concept Development*. Van Haren Publishing.

Crapo, A. W., Waisel, L. B., Wallace, W. A., & Willemain, T. R. (2000). Visualization and the process of modeling. *Proceedings of the Sixth ACM SIGKDD International Conference on Knowledge Discovery and Data Mining - KDD '00*. https://doi.org/10.1145/347090.347129

Cuoco, A., Paul Goldenberg, E., & Mark, J. (1996). Habits of mind: An organizing principle for mathematics curricula. *The Journal of Mathematical Behavior*, *15*(4), 375–402. https://doi.org/10.1016/s0732-3123(96)90023-1

Darling-Hammond, L. (2016, July 28). *Unequal Opportunity: Race and Education*. Brookings. https://www.brookings.edu/articles/unequal-opportunity-race-and-education/

Edblad, P. (2021, May 20). *If-Then Planning: How to Make Good Habits Stick*. Patrik

Edblad. https://patrikedblad.com/habits/if-then-planning/

Fernandes, M. A., Wammes, J. D., & Meade, M. E. (2018). The Surprisingly Powerful Influence of Drawing on Memory. *Current Directions in Psychological Science, 27*(5), 302–308. https://doi.org/10.1177/0963721418755385

Fry, H. (2019). What Statistics Can and Can't Tell Us About Ourselves. *The New Yorker, 9 September 2019*.

Giganti, P. (n.d.). The Mathematical Art of Guessing. *CMC ComMuniCator, 36*(3).

Gillette, H. (2022, September 14). *Patternicity: What It Means When You See Patterns*. Psych Central. https://psychcentral.com/lib/patterns-the-need-for-order

Globokar, L. (2020, March 5). *The Power Of Visualization And How To Use It*. Forbes. https://www.forbes.com/sites/lidijaglobokar/2020/03/05/the-power-of-visualization-and-how-to-use-it/?sh=7173f3e46497

How Math Changed The World One Equation at a Time | Daily Infographic. (2021, July 2). Daily Infographic | Learn Something New Every Day.

https://www.dailyinfographic.com/mathematicians-history

Imagery: A Key to Understanding Math. (2013, October 31). KQED. https://www.kqed.org/mindshift/32208/imagery-a-key-to-understanding-math

Japanese Numbers: How to Count from 1 to 100. (n.d.). Busuu. https://www.busuu.com/en/japanese/numbers

Karp, K. S., Bush, S. B., & Dougherty, B. J. (2014). 13 Rules That Expire. *Teaching Children Mathematics*, *21*(1), 18–25. https://doi.org/10.5951/teacchilmath.21.1.0018

Kaya, A. (2019). The Mathematical Patterns Around Us. *The Fountain*, *129*.

Kayne, D. (2014, October 3). *Number Talks*. Five Twelve Thirteen. Retrieved December 5, 2022, from https://fivetwelvethirteen.wordpress.com/category/number-talks/

Language Magazine. (2013, April 8). *Mathematically Speaking*. https://www.languagemagazine.com/mathematically-speaking/

Lockhart, P. (2009). *A mathematician's lament* (1st ed.). Bellevue Literary Press.

Looney, S. (2022). *Same But Different Math* (1st ed.). Routledge.

Lucky Sevens Film Challenge - Las Vegas 2021. (n.d.). Indiegogo. https://www.indiegogo.com/projects/lucky-sevens-film-challenge-las-vegas-2021

Math Should Be A Gateway. (2021, August 23). https://usprogram.gatesfoundation.org/news-and-insights/articles/what-were-learning-math-should-be-a-gateway

Melina, R. (2011, February 17). *Why Do Most Prices End in .99?* livescience.com. https://www.livescience.com/33045-why-do-most-prices-end-in-99-cents-.html

National Governors Association Center for Best Practices & Council of Chief State School

NOAA predicts above-normal 2022 Atlantic Hurricane Season. (2022, May 24). National Oceanic and Atmospheric Administration. https://www.noaa.gov/news-release/noaa-predicts-above-normal-2022-atlantic-hurricane-season

Officers. (2010). *Common Core State Standards for Mathematics.* Washington, DC:

pattern_1 noun - Definition, pictures, pronunciation and usage notes | Oxford Advanced Learner's Dictionary at OxfordLearnersDictionaries.com. (n.d.-b).

Retrieved November 10, 2022, from https://www.oxfordlearnersdictionaries.com/us/definition/english/pattern_1

Pym, O. (2017, June 13). *Habit Stacking: How To Train Your Brain With Routine*. Esquire. https://www.esquire.com/uk/life/fitness-wellbeing/a15489/habit-stacking-chaining/

Ramajunan, the Man who Saw the Number Pi in Dreams. (n.d.). Retrieved October 27, 2022, from https://www.bbvaopenmind.com/en/science/leading-figures/ramanujan-the-man-who-saw-the-number-pi-in-dreams/

Ranganathan, V. K., Siemionow, V., Liu, J. Z., Sahgal, V., & Yue, G. H. (2004). From mental power to muscle power—gaining strength by using the mind. *Neuropsychologia, 42*(7), 944–956. https://doi.org/10.1016/j.neuropsychologia.2003.11.018

Responding to critics of Common Core math in the elementary grades. (n.d.). The Thomas B. Fordham Institute. https://fordhaminstitute.org/national/commentary/responding-critics-common-core-math-elementary-grades

Rich, E. (2010, October 11). How do you define 21st-century learning?: One question. Eleven answers. *Education Week*.

https://www.edweek.org/teaching-learning/how-do-you-define-21st-century-learning/2010/10?s_kwcid=AL!6416!3!612903424477!!!g!!&utm_source=goog&utm_medium=cpc&utm_campaign=ew+dynamic+ads&ccid=dynamic+ads&ccag=networking+dynamic&cckw=&cccv=dynamic+ad&gclid=Cj0KCQjw--2aBhD5ARIsALiRlwB5tkus_L5aTGZnnhrGNiVwwd88VgtTNzQdEoY_IP_y22sjh6dEz0gaApRaEALw_wcB

Robinson, K. (2008, June 16). *Changing education paradigms*. RSA Events. London, England. https://www.ted.com/talks/sir_ken_robinson_changing_education_paradigms

Rochead. (n.d.). A Good Mathematician is a Lazy Mathematician. In *Institute of Mathematics* [Lecture]. https://ima.org.uk/12527/a-good-mathematician-is-a-lazy-mathematician/

Rockmore, D. (2019, November 7). *The Myth and Magic of Generating New Ideas*. The New Yorker. https://www.newyorker.com/culture/annals-of-inquiry/the-myth-and-magic-of-generating-new-ideas

Roos, D. (2022, September 14). *Thanks, Math! Four Times Numbers Changed the World.*

HowStuffWorks. https://science.howstuffworks.com/math-concepts/math-changed-world.htm

Saffran, J. R., Aslin, R. N., & Newport, E. L. (1996). Statistical Learning by 8-Month-Old Infants. *Science*, *274*(5294), 1926–1928. https://doi.org/10.1126/science.274.5294.1926

Serino, L. (2022, March 9). *What international test scores reveal about American education*. Brookings. https://www.brookings.edu/blog/brown-center-chalkboard/2017/04/07/what-international-test-scores-reveal-about-american-education/

Silver, C. (2019, July 8). *The Mighty Deeds of King Shulgi of Ur, Master of Mesopotamian Monarchs*. Ancient Origins Reconstructing the Story of Humanity's Past. https://www.ancient-origins.net/history-famous-people/king-shulgi-0011602

Sparks, S. D. (2021, February 2). *Number Sense, Not Counting Skills, Predicts Math Ability, Says Study*. Education Week. https://www.edweek.org/education/number-sense-not-counting-skills-predicts-math-ability-says-study/2013/02

Szalay, J. (2017, September 18). *Who Invented Zero?* livescience.com. https://www.livescience.com/27853-who-invented-zero.html

The Applications of Calculus in Everyday Life (Uses & Examples). (2022, June 20). BYJU'S Future School Blog. https://www.byjusfutureschool.com/blog/the-application-of-calculus-in-everyday-life/

The archive. (n.d.). *What If?* https://what-if.xkcd.com

The Editors of Encyclopaedia Britannica. (1998, July 20). *quadrature | mathematics.* Encyclopedia Britannica. https://www.britannica.com/science/quadrature-mathematics

The Harvard Educational Review - HEPG. (n.d.). https://www.hepg.org/her-home/issues/harvard-educational-review-volume-73-issue-4/herbooknote/classroom-discourse_75

The Lottery: Is It Ever Worth Playing? (2022, July 28). Investopedia. https://www.investopedia.com/managing-wealth/worth-playing-lottery/

The Science of Extreme Weather (season 1, episode 51). (2022, November 11). NPR.

https://www.npr.org/podcasts/381443461/the-pulse

The Story of Mathematics - A History of Mathematical Thought from Ancient Times to the Modern Day. (2022, January 11). *Sumerian/Babylonian Mathematics*. The Story of Mathematics - a History of Mathematical Thought From Ancient Times to the Modern Day. https://www.storyofmathematics.com/sumerian.html/

Thinking Through, and By, Visualising. (n.d.). https://nrich.maths.org/6447

This is your brain on mathematics. (2016, April 20). The Intrepid Mathematician. https://anthonybonato.com/2016/04/20/this-is-your-brain-on-mathematics/

Timotheou, S., & Ioannou, A. (n.d.). *On a Making-&-Tinkering Approach to Learning Mathematics in Formal Education: Knowledge Gains, Attitudes, and 21st-Century Skills* [Conference paper]. CSCL Conference 2019. https://www.researchgate.net/publication/342663033_On_a_Making--Tinkering_Approach_to_Learning_Mathematics_in_Formal_Education_Knowledge_Gains_Attitudes_and_21_st_-Century_Skills

Tuarez, J. (2021, February 22). *How do mathematicians think?* NeuroTray. https://neurotray.com/how-do-mathematicians-think/

Volpe, M. (2013, March 21). *5 Shocking Statistics - How Junk Mail Marketing Damages the Environment.* https://blog.hubspot.com/blog/tabid/6307/bid/3741/5-shocking-statistics-how-junk-mail-marketing-damages-the-environment.aspx

What is the origin of zero? How did we indicate nothingness before zero? (2007, January 16). Scientific American. https://www.scientificamerican.com/article/what-is-the-origin-of-zer/

Wikipedia contributors. (2022, November 14). *Fibonacci number.* Wikipedia. https://en.wikipedia.org/wiki/Fibonacci_number

Wikipedia contributors. (2022, October 22). *Shulgi.* Wikipedia. https://en.wikipedia.org/wiki/Shulgi

Zakarin, J. (2020, September 8). *Paul McCartney came up with the melody to one of the Beatles' biggest hits in his sleep.* Biography. https://www.biography.com/news/paul-mccartney-the-beatles-yesterday-dream

Zimmerman, A. (2015). *A Guide for Parents to Non-Standard Methods for Multi-Digit Multiplication*. Scholastic. https://www.scholastic.com/content/dam/teachers/blogs/alycia-zimmerman/migrated-files/non-standard_multiplication_methods_parents_letter.pdf

Endnotes

---

[i] Lockhart, P. (2009). *A mathematician's lament* (1st ed.). Bellevue Literary Press.

[ii] Zakarin, J. (2020, September 8). *Paul McCartney came up with the melody to one of the Beatles' biggest hits in his sleep.* Biography. https://www.biography.com/news/paul-mccartney-the-beatles-yesterday-dream

[iii] *Ramajunan, the Man who Saw the Number Pi in Dreams.* (n.d.). Retrieved October 27, 2022, from https://www.bbvaopenmind.com/en/science/leading-figures/ramanujan-the-man-who-

saw-the-number-pi-in-dreams/

[iv] Rockmore, D. (2019, November 7). *The Myth and Magic of Generating New Ideas*. The New Yorker. https://www.newyorker.com/culture/annals-of-inquiry/the-myth-and-magic-of-generating-new-ideas

[v] *This is your brain on mathematics*. (2016, April 20). The Intrepid Mathematician. https://anthonybonato.com/2016/04/20/this-is-your-brain-on-mathematics/

[vi] Tuarez, J. (2021, February 22). *How do mathematicians think?* NeuroTray. https://neurotray.com/how-do-mathematicians-think/

[vii] Tuarez, J. (2021, February 22). *How do mathematicians think?* NeuroTray. https://neurotray.com/how-do-mathematicians-think/

[viii] Cuoco, A., Paul Goldenberg, E., & Mark, J. (1996). Habits of mind: An organizing principle for mathematics curricula. *The Journal of Mathematical Behavior, 15*(4), 375–402. https://doi.org/10.1016/s0732-3123(96)90023-1

[ix] Cuoco, A., Paul Goldenberg, E., & Mark, J. (1996). Habits of mind: An organizing principle for mathematics curricula. *The Journal of Mathematical Behavior, 15*(4), 375–402. https://doi.org/10.1016/s0732-

3123(96)90023-1

[x] National Governors Association Center for Best Practices & Council of Chief State School Officers. (2010). *Common Core State Standards for Mathematics.* Washington, DC: Authors.

[xi] Rich, E. (2010, October 11). How do you define 21st-century learning?: One question. Eleven answers. *Education Week.* https://www.edweek.org/teaching-learning/how-do-you-define-21st-century-learning/2010/10?s_kwcid=AL!6416!3!612903424477!!!g!!&utm_source=goog&utm_medium=cpc&utm_campaign=ew+dynamic+ads&ccid=dynamic+ads&ccag=networking+dynamic&cckw=&cccv=dynamic+

ad&gclid=Cj0KCQjw--

2aBhD5ARIsALiRlwB5tkus_L5aTGZnnhr

GNiVwwd88VgtTNzQdEoY_IP_y22sjh6d

Ez0gaApRaEALw_wcB

[xii] Robinson, K. (2008, June 16). *Changing education paradigms*. RSA Events. London, England. https://www.ted.com/talks/sir_ken_robinson_changing_education_paradigms

[xiii] Robinson, K. (2008, June 16). *Changing education paradigms*. RSA Events. London, England. https://www.ted.com/talks/sir_ken_robinson_changing_education_paradigms

[xiv] Gillette, H. (2022, September 14). *Patternicity:*

What It Means When You See Patterns. Psych Central. https://psychcentral.com/lib/patterns-the-need-for-order

[xv] Barkman, PhD, R. C. (2021, May 19). Why the Human Brain is So Good at Detecting Patterns. *Psychology Today: A Singular Perspective*. https://www.psychologytoday.com/us/blog/singular-perspective/202105/why-the-human-brain-is-so-good-detecting-patterns

[xvi] Saffran, J. R., Aslin, R. N., & Newport, E. L. (1996). Statistical Learning by 8-Month-Old Infants. *Science*, *274*(5294), 1926–1928.

https://doi.org/10.1126/science.274.5294.1926

[xvii] *pattern_1 noun - Definition, pictures, pronunciation and usage notes | Oxford Advanced Learner's Dictionary at OxfordLearnersDictionaries.com.* (n.d.-b). Retrieved November 10, 2022, from https://www.oxfordlearnersdictionaries.com/us/definition/english/pattern_1

[xviii] *The Science of Extreme Weather* (season 1, episode 51). (2022, November 11). NPR. https://www.npr.org/podcasts/381443461/the-pulse

[xix] Wikipedia contributors. (2022, November 14). *Fibonacci number*. Wikipedia.

https://en.wikipedia.org/wiki/Fibonacci_number

[xx] Kaya, A. (2019). The Mathematical Patterns Around Us. *The Fountain, 129.*

[xxi] *Lucky Sevens Film Challenge - Las Vegas 2021.* (n.d.). Indiegogo. https://www.indiegogo.com/projects/lucky-sevens-film-challenge-las-vegas-2021

[xxii] Fry, H. (2019). What Statistics Can and Can't Tell Us About Ourselves. *The New Yorker, 9 September 2019.*

[xxiii] *The Lottery: Is It Ever Worth Playing?* (2022, July 28). Investopedia. https://www.investopedia.com/managing-

wealth/worth-playing-lottery/

[xxiv] *Japanese Numbers: How to Count from 1 to 100.* (n.d.). Busuu. https://www.busuu.com/en/japanese/numbers

[xxv] Language Magazine. (2013, April 8). *Mathematically Speaking.* https://www.languagemagazine.com/mathematically-speaking/

[xxvi] *Math Should Be A Gateway.* (2021, August 23). https://usprogram.gatesfoundation.org/news-and-insights/articles/what-were-learning-math-should-be-a-gateway

[xxvii] *NOAA predicts above-normal 2022 Atlantic Hurricane Season.* (2022, May 24).

National Oceanic and Atmospheric Administration. https://www.noaa.gov/news-release/noaa-predicts-above-normal-2022-atlantic-hurricane-season

[xxviii] *The Harvard Educational Review - HEPG.* (n.d.). https://www.hepg.org/her-home/issues/harvard-educational-review-volume-73-issue-4/herbooknote/classroom-discourse_75

[xxix] A Discourse Primer for Science Teachers. (2015). In *Ambitious Science Teaching.*

[xxx] Timotheou, S., & Ioannou, A. (n.d.). *On a Making-&-Tinkering Approach to Learning Mathematics in Formal*

Education: Knowledge Gains, Attitudes, and 21st-Century Skills [Conference paper]. CSCL Conference 2019. https://www.researchgate.net/publication/342663033_On_a_Making--Tinkering_Approach_to_Learning_Mathematics_in_Formal_Education_Knowledge_Gains_Attitudes_and_21_st_-Century_Skills

[xxxi] Sparks, S. D. (2021, February 2). *Number Sense, Not Counting Skills, Predicts Math Ability, Says Study*. Education Week. https://www.edweek.org/education/number-sense-not-counting-skills-predicts-math-ability-says-study/2013/02

[xxxii] The archive. (n.d.). *What If?* https://what-if.xkcd.com

[xxxiii] Cardone, T. (2015). *Nix the Tricks: A Guide to Avoiding Shortcuts That Cut Out Math Concept Development*. Van Haren Publishing.

[xxxiv] Karp, K. S., Bush, S. B., & Dougherty, B. J. (2014). 13 Rules That Expire. *Teaching Children Mathematics*, *21*(1), 18–25. https://doi.org/10.5951/teacchilmath.21.1.0018

[xxxv] *Responding to critics of Common Core math in the elementary grades*. (n.d.). The Thomas B. Fordham Institute. https://fordhaminstitute.org/national/comm

entary/responding-critics-common-core-math-elementary-grades

[xxxvi] Darling-Hammond, L. (2016, July 28). *Unequal Opportunity: Race and Education*. Brookings. https://www.brookings.edu/articles/unequal-opportunity-race-and-education/

[xxxvii] Serino, L. (2022, March 9). *What international test scores reveal about American education*. Brookings. https://www.brookings.edu/blog/brown-center-chalkboard/2017/04/07/what-international-test-scores-reveal-about-american-education/

[xxxviii] Zimmerman, A. (2015). *A Guide for Parents*

to Non-Standard Methods for Multi-Digit Multiplication. Scholastic. https://www.scholastic.com/content/dam/teachers/blogs/alycia-zimmerman/migrated-files/non-standard_multiplication_methods_parents_letter.pdf

[xxxix] *Algorithms in Mathematics and Beyond*. (2018, December 21). ThoughtCo. https://www.thoughtco.com/definition-of-algorithm-2312354

[xl] Rochead. (n.d.). A Good Mathematician is a Lazy Mathematician. In *Institute of Mathematics* [Lecture]. https://ima.org.uk/12527/a-good-

mathematician-is-a-lazy-mathematician/

[xli] Edblad, P. (2021, May 20). *If-Then Planning: How to Make Good Habits Stick*. Patrik Edblad. https://patrikedblad.com/habits/if-then-planning/

[xlii] Edblad, P. (2021, May 20). *If-Then Planning: How to Make Good Habits Stick*. Patrik Edblad. https://patrikedblad.com/habits/if-then-planning/

[xliii] Pym, O. (2017, June 13). *Habit Stacking: How To Train Your Brain With Routine*. Esquire. https://www.esquire.com/uk/life/fitness-wellbeing/a15489/habit-stacking-chaining/

[xliv] Volpe, M. (2013, March 21). *5 Shocking Statistics - How Junk Mail Marketing Damages the Environment.* https://blog.hubspot.com/blog/tabid/6307/bid/3741/5-shocking-statistics-how-junk-mail-marketing-damages-the-environment.aspx

[xlv] Crapo, A. W., Waisel, L. B., Wallace, W. A., & Willemain, T. R. (2000). Visualization and the process of modeling. *Proceedings of the Sixth ACM SIGKDD International Conference on Knowledge Discovery and Data Mining - KDD '00.* https://doi.org/10.1145/347090.347129

[xlvi] *Imagery: A Key to Understanding Math.* (2013,

October 31). KQED. https://www.kqed.org/mindshift/32208/imagery-a-key-to-understanding-math

[xlvii] *Thinking Through, and By, Visualising.* (n.d.). https://nrich.maths.org/6447

[xlviii] Fernandes, M. A., Wammes, J. D., & Meade, M. E. (2018). The Surprisingly Powerful Influence of Drawing on Memory. *Current Directions in Psychological Science*, *27*(5), 302–308. https://doi.org/10.1177/0963721418755385

[xlix] *Achievethecore.org :* (n.d.). https://achievethecore.org/content/upload/Mathematical+Routines.pdf

[l] Looney, S. (2022). *Same But Different Math* (1st ed.). Routledge.

[li] Globokar, L. (2020, March 5). *The Power Of Visualization And How To Use It*. Forbes. https://www.forbes.com/sites/lidijagloboka r/2020/03/05/the-power-of-visualization-and-how-to-use-it/?sh=7173f3e46497

[lii] Ranganathan, V. K., Siemionow, V., Liu, J. Z., Sahgal, V., & Yue, G. H. (2004). From mental power to muscle power—gaining strength by using the mind. *Neuropsychologia, 42*(7), 944–956. https://doi.org/10.1016/j.neuropsychologia.2003.11.018

[liii] Giganti, P. (n.d.). The Mathematical Art of

Guessing. *CMC ComMuniCator*, *36*(3).

[liv] Melina, R. (2011, February 17). *Why Do Most Prices End in .99?* livescience.com. https://www.livescience.com/33045-why-do-most-prices-end-in-99-cents-.html

[lv] Kayne, D. (2014, October 3). *Number Talks*. Five Twelve Thirteen. Retrieved December 5, 2022, from https://fivetwelvethirteen.wordpress.com/category/number-talks/

[lvi] Wikipedia contributors. (2022, October 22). *Shulgi*. Wikipedia. https://en.wikipedia.org/wiki/Shulgi

[lvii] Silver, C. (2019, July 8). *The Mighty Deeds of*

King Shulgi of Ur, Master of Mesopotamian Monarchs. Ancient Origins Reconstructing the Story of Humanity's Past. https://www.ancient-origins.net/history-famous-people/king-shulgi-0011602

[lviii] The Story of Mathematics - A History of Mathematical Thought from Ancient Times to the Modern Day. (2022, January 11). *Sumerian/Babylonian Mathematics*. The Story of Mathematics - a History of Mathematical Thought From Ancient Times to the Modern Day. https://www.storyofmathematics.com/sumerian.html/

[lix] The Story of Mathematics - A History of Mathematical Thought from Ancient Times to the Modern Day. (2022, January 11). *Sumerian/Babylonian Mathematics*. The Story of Mathematics - a History of Mathematical Thought From Ancient Times to the Modern Day. https://www.storyofmathematics.com/sumerian.html/

[lx] Szalay, J. (2017, September 18). *Who Invented Zero?* livescience.com. https://www.livescience.com/27853-who-invented-zero.html

[lxi] *What is the origin of zero? How did we indicate nothingness before zero?* (2007, January

16). Scientific American. https://www.scientificamerican.com/article/what-is-the-origin-of-zer/

[lxii] The Editors of Encyclopaedia Britannica. (1998, July 20). *quadrature | mathematics*. Encyclopedia Britannica. https://www.britannica.com/science/quadrature-mathematics

[lxiii] Al Jazeera. (2015, October 20). *Al-Khwarizmi: The Father of Algebra*. Science and Technology | Al Jazeera. https://www.aljazeera.com/program/science-in-a-golden-age/2015/10/20/al-khwarizmi-the-father-of-algebra

[lxiv] Roos, D. (2022, September 14). *Thanks, Math!*

Four Times Numbers Changed the World. HowStuffWorks. https://science.howstuffworks.com/math-concepts/math-changed-world.htm

[lxv] *The Applications of Calculus in Everyday Life (Uses & Examples).* (2022, June 20). BYJU'S Future School Blog. https://www.byjusfutureschool.com/blog/the-application-of-calculus-in-everyday-life/

[lxvi] Buell, B. (2017, June 18). 2 wizards kept GE atop technology field. *The Daily Gazette.* https://dailygazette.com/2017/06/17/edison-steinmetz-forged-long-friendship/

[lxvii] Roos, D. (2022, September 14). *Thanks, Math! Four Times Numbers Changed the World.* HowStuffWorks. https://science.howstuffworks.com/math-concepts/math-changed-world.htm

[lxviii] *How Math Changed The World One Equation at a Time | Daily Infographic.* (2021, July 2). Daily Infographic | Learn Something New Every Day. https://www.dailyinfographic.com/mathematicians-history